WILD URBAN PLANTS
OF THE NORTHEAST

Peter Del Tredici

Wild Urban Plants
OF THE *Northeast*

A FIELD GUIDE

SECOND EDITION

Foreword by Steward T. A. Pickett

Comstock Publishing Associates   *an imprint of*
Cornell University Press   ITHACA *&* LONDON

First edition first published 2010 by Cornell University Press
Second edition first published 2020 by Cornell University Press
Printed in China

LIBRARY OF CONGRESS CATALOGING-IN-PUBLICATION DATA
Names: Del Tredici, Peter, 1945– author.
Title: Wild urban plants of the Northeast : a field guide / Peter Del Tredici ;
foreword by Steward T.A. Pickett.
Description: Second edition. | Ithaca : Comstock Publishing Associates,
an imprint of Cornell University Press, 2020. | Includes bibliographical
references and index.
Identifiers: LCCN 2019025507 (print) | LCCN 2019025508 (ebook) |
ISBN 9781501740442 (paperback) | ISBN 9781501740459 (pdf) |
ISBN 9781501740466 (epub)
Subjects: LCSH: Urban plants—Northeastern States—Identification. |
Weeds—Northeastern States—Identification.
Classification: LCC QK118 .D45 2020 (print) | LCC QK118 (ebook) |
DDC 581.7/560974—dc23
LC record available at https://lccn.loc.gov/2019025507
LC ebook record available at https://lccn.loc.gov/2019025508

To my grandchildren
for whom the ordinary
is extraordinary

CONTENTS

Foreword to the Second Edition *by Steward T. A. Pickett, xi*

Preface to the Second Edition, *xiii*

Acknowledgments and Photography Credits, *xv*

Introduction, *1*

MOSSES AND FERNS

Bryaceae (Byrum Moss Family), *30*

Dryopteridaceae (Woodfern Family), *32*

HORSETAILS

Equisetaceae (Horsetail Family), *34*

CONIFERS

Taxaceae (Yew Family), *36*

WOODY DICOTS

Anacardiaceae (Cashew Family), *38*

Berberidaceae (Barberry Family), *42*

Betulaceae (Birch Family), *44*

Bignoniaceae (Trumpet Creeper Family), *50*

Cannabaceae (Hemp Family), *52*

Caprifoliaceae (Honeysuckle Family), *54*

Celastraceae (Stafftree Family), *58*

Elaeagnaceae (Oleaster Family), *62*

Fabaceae = Leguminosae (Pea Family), *64*

Fagaceae (Beech Family), *70*

Juglandaceae (Walnut Family), *74*

Moraceae (Mulberry Family), *76*

Oleaceae (Olive Family), *78*

Paulowniaceae (Princess Tree Family), *82*

Ranunculaceae (Buttercup Family), *84*

Rhamnaceae (Buckthorn Family), *86*

Rosaceae (Rose Family), *90*

Rutaceae (Rue Family), *104*

Salicaceae (Willow Family), *106*

Sapindaceae (Soapwort Family), *112*

Simaroubaceae (Quassia Family), *120*
Solanaceae (Nightshade Family), *122*
Ulmaceae (Elm Family), *124*
Vitaceae (Grape Family), *128*

Herbaceous Dicots

Amaranthaceae (Amaranth Family), *134*
Apiaceae = Umbelliferae (Carrot Family), *138*
Apocynaceae (Dogbane Family), *140*
Asclepiadaceae (Milkweed Family), *142*
Asteraceae = Compositae (Aster Family), *146*
Balsaminaceae (Touch-me-not Family), *196*
Brassicaceae = Cruciferae (Mustard Family), *198*
Cannabaceae (Hemp Family), *210*
Caryophyllaceae (Pink Family), *212*
Convolvulaceae (Morning Glory Family), *224*
Crassulaceae (Stonecrop Family), *226*
Cucurbitaceae (Gourd Family), *228*
Euphorbiaceae (Spurge Family), *230*
Fabaceae = Leguminosae (Pea Family), *236*
Hyperiaceae (St. John's Wort Family), *250*
Lamiaceae = Labiatae (Mint Family), *252*
Lythraceae (Loosestrife Family), *260*
Malvaceae (Mallow Family), *262*
Molluginaceae (Carpetweed Family), *264*
Onagraceae (Evening Primrose Family), *266*
Oxalidaceae (Woodsorrel Family), *268*
Papaveraceae (Poppy Family), *270*
Phytolaccaceae (Pokeweed Family), *272*
Plantaginaceae (Plantain Family), *274*
Polygonaceae (Smartweed Family), *282*
Portulacaceae (Purslane Family), *296*
Ranunculaceae (Buttercup Family), *298*
Rosaceae (Rose Family), *302*
Rubiaceae (Madder Family), *306*
Scrophulariaceae (Figwort Family), *308*
Solanaceae (Nightshade Family), *310*
Urticaceae (Nettle Family), *314*
Verbenaceae (Verbena Family), *316*
Violaceae (Violet Family), *318*

Monocots

Amaryllidaceae (Amaryllis Family), *320*
Commelinaceae (Spiderwort Family), *322*
Cyperaceae (Sedge Family), *324*

Iridaceae (Iris Family), *326*
Juncaceae (Rush Family), *328*
Poaceae = Gramineae (Grass Family), *330*
Smilacaceae (Smilax Family), *362*
Typhaceae (Cattail Family), *364*

APPENDIXES

1. Urban Habitats and Their Preadapted Plants, *367*
2. Plants Treated in This Book That Are Included in Dioscorides'
   De Materia Medica, *370*
3. European Plants Listed by John Josselyn as Growing Spontaneously
   in New England in the Seventeenth Century, *371*
4. Species Suitable for a Cosmopolitan Urban Meadow, *372*
5. Shade-Tolerance Ratings of the 40 Trees Covered in This Book, *373*
6. Key Characteristics of Important Plant Families, *375*

Glossary, *379*

References, *391*

Index, *399*

FOREWORD TO THE SECOND EDITION

Ten years ago, Peter Del Tredici produced a very useful and thoughtful field guide to the common urban plants of the northeastern United States. The region includes some of our largest urban agglomerations and suburban territories, extending up and down the East Coast from Montreal to Washington, D.C., and from Boston to Detroit. That first edition provided an important resource for urban residents and land managers by giving them a convenient tool to identify and learn about the plants that inhabit their cities. It is a welcome event to mark the publication of a new edition of that pioneering book that adds forty-five new species to the plant roster—an increase of 20% over its predecessor.

Urban ecologists and climate scientists have become more convinced that the continued human movement of organisms to cities, along with changes in climate and associated local environmental conditions, have transformed them into "novel ecosystems." These new combinations of species have no analogue in the past and their future trajectories are unknown. On the one hand, this realization suggests that the plants that spontaneously establish in cities are capable of adapting to on-going environmental change; on the other hand, it shows that managing the ecology of urban ecosystems strictly by looking backward is a poor strategy for keeping pace with our rapidly changing environment.

These conclusions may seem radical and, to some perhaps, disturbing. The new introduction to *Wild Urban Plants of the Northeast*, however, provides a concise and easily comprehensible summary of the key lessons of modern urban ecology that support these conclusions. It includes well-documented and thoughtful descriptions of the origins, characteristics, evolution, and benefits of plants that establish spontaneously in urban habitats. This introduction is, in fact, a very good survey in approachable language of relevant findings from modern urban ecology.

Readers of this book will garner much more understanding than simply knowing what plants are growing in their neighborhood or in the neglected slivers of land in their cities or towns. Readers will begin to appreciate more fully the role of ecology—the patterns and processes influenced by organisms of all sorts—in the city. They will also come to appreciate how much disturbance and intentional change takes place in cities, and how plants respond to these pressures. And finally, they will come to understand the immense amount of "free work" that spontaneous plants perform in urban systems. With this understanding, they will question the simplistic and judgmental focus on species labeled as exotic or invasive, and instead, will evaluate the resulting novel ecosystems by the functions and services they provide.

This book provides readers with a viable alternative to the usual simplistic labels of species as either native or non-native. Instead, they will find a rich section on the "Cultural Significance" of each species that illuminates the compelling connections between humans and the plants that surround them in the urban environment—many of them dating back thousands of years. These human–plant interactions are continuing to evolve along with our changing climate, and readers will see how exciting an arena the cities are for understanding the ongoing process of plant evolution in an urbanized world. Specialists in ecological research who delve deeply into the book will find some very nice ideas to stimulate new research.

This new edition is important to a broad audience who already have an interest in plants, but it is also an important statement of the emerging and changing ecological reality of our cities. *Wild Urban Plants of the Northeast* is not just an updated tool for the task of identifying individual species of plants in cities and towns and appreciating their history. It is also an important tool to help land managers, policymakers, landscape architects, garden designers, maintenance contractors, and residents better understand, and thus better manage, novel urban ecosystems. Few tasks are more important for people who inhabit our increasingly urbanized world.

Steward T. A. Pickett
Cary Institute of Ecosystem Studies *and*
Baltimore Ecosystem Study
Millbrook, New York

PREFACE TO THE SECOND EDITION

The genesis of this book goes back sixteen years—to the spring of 2004—when I was on a field trip with my landscape architecture students from the Harvard Graduate School of Design. We were visiting Spectacle Island, a capped landfill in the middle of Boston Harbor, to study the vegetation. At some point, as I was explaining how to identify a particular weed growing alongside the pathway, I casually mentioned how unfortunate it was that we lacked the class time to learn about the weedy plants that are so common in the urban environment. The ecology course covered native species, and my class dealt with cultivated species, but nobody talked about the weeds that were growing everywhere. The curiosity of one of the students, Leah Broder, was piqued by this comment, and she immediately suggested that maybe we could set up a website for the class that would help students learn to identify these plants on their own time. Without giving it much thought, I agreed and naively suggested that with her help maybe it could be done in time for next spring's class and that we could use slides I already had in my collection. Leah immediately agreed to help with the project.

Back at the Design School, I quickly put together a grant proposal for the Harvard Center for Innovative Teaching Technologies, and they quickly approved it. With money in hand and technical support from the Design School's IT specialist Kevin Lau, I was able to hire and train Leah and another of my students, Ken Francis, to begin working on the project. In August 2005—some sixteen months later—we launched the first version of the *Emergent Vegetation of the Urban Environment* (EVUE) website in time for the fall semester.

With a continuation of the grant for a second year and additional help from two new students, Sharon Komarow and Addie Pierce McManamon, we expanded the website to nearly one hundred species and added digital images to replace blurry scanned slides. Shortly after the launch of EVUE 2.0 in September 2006, the money from Harvard ran out, leaving the website fully functional but frozen in time. It was at this point that I approached Cornell University Press about putting the information and photos from the website into book form and expanding the number of species covered. Producing the book version of the website took an additional four years, and the first edition of *Wild Urban Plants of the Northeast* came out in spring 2010.

Now, ten years later, you hold the second edition of this book in your hands. It covers forty-five new plants (a 20% increase), which brings the total coverage to 268 species. Some of these additions are plants that have become more common in urban areas since the first edition came out, while others probably should have been included in the first edition but were left out because of space limitations or a lack of data about the extent of their distribution. In addition, the introduction to the second edition has been totally rewritten to reflect the increasingly obvious impacts of climate change on urban ecology. Conditions that were once seen as looming in the future have now been recognized as the "new normal" for cities across the globe. Indeed, most of the plants described in this book, when confronted with resources that have become more available as a result of urbanization—CO_2, soil nitrogen, and heat—respond by growing more rapidly and reproducing more abundantly. Many of our native forest species, by contrast, are adapted to environments with scarce resources and are at a loss for how to handle the ten pounds of nitrogen per acre that is falling from the sky every year as a result of the burning of fossil fuels. Regardless of how we feel about it, the world of tomorrow belongs to those plants that can keep pace with rapidly changing environmental conditions. In this sense, *Wild Urban Plants of the Northeast* is a field guide to flora of the future.

ACKNOWLEDGMENTS

I am grateful to many people for their assistance in the production of this book. With the first edition, I am indebted to Randy Prostak of the University of Massachusetts, who reviewed the manuscript and photos for accuracy; Les Mehrhoff of the University of Connecticut (now deceased), who helped me with the identification of several problematic species; and Roxana and Ledlie Laughlin of Cornwall, Connecticut, who generously allowed me to use their beautiful cabin in the woods to complete the manuscript during the summer of 2008. I also want to acknowledge the support of my two children, Sonya and Luke, who uncomplainingly put up with innumerable travel delays while I took yet another "weed" picture, and of my wife, Susan Klaw, for her willingness not only to make sudden stops by the side of the road to let me take photos with her iPhone but also to stand in the pictures for scale. I cannot overstate the importance of her encouragement and patience during the sixteen years I have been working on this book.

For the second edition, I would like to acknowledge the encouragement and assistance of professional colleagues and general readers who pointed out errors in the first edition, suggested additions to the species treatments, and/or offered constructive criticism on the content of the book. Because of this input, the second edition has 20% more plant entries than the first edition and is considerably more accurate. Note that I used the opportunity of a second edition to replace photographs that were too small to be legible with images that worked better in the available space. Finally, I acknowledge the support of my editor at Cornell University Press, Kitty Liu, who shepherded this book through the editorial process.

Photography Credits

In addition to the photos provided by **Les Mehrhoff** (*Acer pseudoplatanus* and *Berberis thunbergii* flowers) and **Lou Wagner** (*Vintoxicum rossicum* flowers) for the first edition of this book, the following individuals have generously given me permission to use their excellent photographs in the second edition of this book:

Randall G. Prostak, University of Massachusetts, provided ten images from the Weed Herbarium website he manages at http://extension.umass.edu/landscape/weed-herbarium: *Berteroa incana*, whole plant and flower close-up; *Capsella bursa-pastoris*, mature fruits; *Cardamine hirsuta*, in flower; *Erophila verna*, whole plant;

Euphorbia maculata, flower close-up; *Gallium aparine*, fruits; *Medicago lupulina*, inflorescence; *Potentilla norvegica*, flower; *Scleranthus annuus*, flower close-up and plant in hand; *Verbena urticifolia*, flower close-up.

Will Cook provided the image of *Smilax rotundifolia* flowers.

Thomas H. Kent, FloraFinder.org, provided the image of *Typha angustifolia* flowers.

Robert Klips provided the image of *Bryum argenteum* spore capsules.

Justin Thomas provided the image of *Ceratadon purpurescens* spore capsules.

All other photographs are my own (see more information at the end of the introduction).

WILD URBAN PLANTS
OF THE NORTHEAST

INTRODUCTION

The flora of today surely differs from that of five hundred or more years ago, due largely to the influence of an increasingly complicated civilization; may it not be of interest to record in detail the ruderals and escapes of to-day as a prophesy of the flora of the not distant future?
—Edgar Anderson and Robert Woodson
The Species of Tradescantia, 1935

Ailanthus, sumac, sunflowers, and other weedy wildflowers colonize the wastelands and forgotten corners of the city and provide, at no cost, many of the same services that the cultivated plant communities do. In urban wastelands, they decorate what would otherwise be a desolate environment. But most city dwellers are blinded to their beauty by a more domesticated aesthetic. An unappreciated and neglected resource, their energy goes unharnessed.
—Anne Whiston Spirn
The Granite Garden, 1984

The basic goal of *Wild Urban Plants* is to help the general reader identify the plants that grow spontaneously in the urban environments of the Northeast and develop an appreciation for the role they play in making our cities more livable. The 268 plants featured in this book fill the vacant spaces between our roads, our homes, and our businesses; take over neglected landscapes; and line the ever-changing shores of streams, rivers, lakes, and oceans. Some of the plants are native to the region and were present before humans drastically altered the landscape; some were brought intentionally or unintentionally by people; and some arrived on their own, dispersed by wind, water, or wild animals. They grow and reproduce in the city without being planted or cared for by people. They are everywhere and yet they are invisible to most people.

Given that cities are human creations and that the original vegetation that once grew there has long since disappeared, one could argue that these spontaneous plants have become the *de facto* native urban flora. Indeed, the basic premise of this book is that the ecology of the city is defined not so much by the cultivated plants

that require ongoing maintenance or the native species that are restricted to protected natural areas, but by the plants that dominate the neglected interstices of the urban environment without human permission. This "wasteland" flora occupies a significant percentage of the open space in many American cities, especially those with faltering economies. If such vegetation is left undisturbed long enough to develop into woodlands, it can provide cities with important social and ecological services at no cost to taxpayers (Riley et al. 2018).

The most well-known example of a "spontaneous" urban plant is *Ailanthus altissima*, or tree-of-heaven, from China. Widely planted in the Northeast in the first half of the nineteenth century, *Ailanthus* was later rejected by urban tree planters as uncouth and weedy. Despite concerted efforts at eradication, the tree

Spontaneous trees, such as this tree-of-heaven (*Ailanthus altissima*), have become significant components of the urban forest.

managed to persist by sprouting from its roots and to spread by scattering its seeds to the wind. The niche it occupies in cities was famously described by Betty Smith in the opening page of her 1943 novel, *A Tree Grows in Brooklyn*: "There's a tree that grows in Brooklyn. Some people call it the Tree of Heaven. No matter where its seed falls, it makes a tree which struggles to reach the sky. It grows in boarded-up lots and out of neglected rubbish heaps. It grows up out of cellar gratings. It is the only tree that grows out of cement. It grows lushly . . . survives without sun, water, and seemingly without earth. It should be considered beautiful except that there are too many of it."

Although it is ubiquitous in the urban landscape, *Ailanthus* is seldom counted in street tree inventories because no one planted it and consequently its contribution to making the city a more livable place goes unrecognized. When Mayor Michael Bloomberg of New York City promised to plant a million trees to fight global warming in 2007, he failed to realize that if the *Ailanthus* trees already growing throughout the city were counted, he would be halfway toward his goal without doing anything. And that, of course, is the larger purpose of this book: to open people's eyes to the ecological reality of our cities and to appreciate it for what it is without passing judgment on it. *Ailanthus* is just as good at sequestering carbon and creating shade as our beloved native species or showy horticultural selections. Indeed, if one were to ask whether our cities would be better or worse without *Ailanthus*, the answer would most certainly be the latter, given that it typically grows where few cultivated trees could survive without maintenance.

Common reed (*Phragmites australis*) dominates this "loading dock wetland" in an abandoned Detroit factory.

There is no denying the fact that many, if not most, of the plants covered in this book suffer from image problems associated with the label "weeds," or the more recent term, "invasive species." From the plant's perspective, invasiveness is just another word for successful reproduction—the ultimate goal of all organisms, including humans. From a utilitarian perspective, a weed is any plant that grows by itself in a place where people do not want it to grow. Calling a plant a weed gives people license to get rid of it; similarly, calling a plant invasive allows them to blame it for ruining the environment.

From the ecological perspective, a weed can be defined as a plant that is adapted to disturbance in all its myriad forms, from bulldozers to acid rain. Weeds are symptoms of environmental degradation—not its cause—and as such they are poised to become increasingly abundant as human-driven climate change relentlessly degrades the world's environment. In this sense, the plants described in this book are the flora of the future. They are the plants that can respond positively to increased levels of carbon dioxide, soil nitrogen, and heat while most of our native forest flora is adapted to habitats where resources are scarce and competition fierce. As Edgar Anderson pointed out in 1952 in his "dump heap" theory on the origins of agriculture, weeds have been inseparably linked to the accumulation of human waste for millennia.

What Is a Weed?

I consulted innumerable books and articles about weeds and invasive species in the process of writing this book, and the one concept that stands out is that a weed is a social rather than a biological construct. In the immortal words of Ralph Waldo Emerson (1879), a weed is "a plant whose virtues have not yet been discovered." At its core, a weed is, quite simply, a plant that people do not like because it is growing

Quaking aspen
(*Populus tremuloides*)
colonizing the roof
of the abandoned
Central Train Depot
in Detroit.

where they do not want it to grow. To put it another way, it is the context in which a plant is growing—not the plant itself—that makes a weed. The Romans, who had names for most things, had no specific word for weed. According to various sources, the closest term they used was *viriditas*, which roughly translates to "greenness"— a purely descriptive term lacking value judgement. In the modern Romance languages, the words for weed are simply a negative descriptor added to the word for an herbaceous plant: *mala hierba* in Spanish, *mauvaise herbe* in French, and *erbaccia* in Italian. In German the word for weed, *unkraut,* carries the same negative connotation.

Most books about weeds focus on species that are problematic in either an *agricultural context*, where the issue of competition with economic crops is the primary concern, or a *landscape context*, where an unsightly plant is growing in a place where people are trying to cultivate something else or do not want anything at all to grow (Salisbury 1961). The term *invasive species* is typically used to describe a plant that displaces desirable native vegetation in natural areas in an *ecological context*. This, of course, raises the question of what constitutes an invasive species in an *urban context*, where humans long ago destroyed most of the original vegetation? By the same token, the concept of a native species in an urban context has little meaning beyond its historical significance. In general, people's negative feelings about spontaneous urban vegetation are either aesthetic—they are seen as ugly or indicators of neglect—or fear-based—they provide cover for illicit human activity or habitat for disease-carrying vermin.

Although the overlap is considerable among the three categories of "weeds" listed above, they can be readily distinguished by the types of landscapes in which they grow. In general, agricultural habitats combine annual soil disturbance with high nutrient levels; ecological habitats are characterized by low levels of both soil

The urban glacier (*left*) leaves a trail of compacted glacial till in its wake.

disturbance and nutrients; and urban habitats experience intermediate levels of soil disturbance and nutrients. While disturbance is an integral part of the ecology of all three types of habitats, they differ from one another in the frequency with which the disturbance impacts their ecological structure. This is really just another way of saying that *succession*—the term used by ecologists to describe the changes in the composition of biological communities over time—is driven by disturbance. The initial stages of the process are referred to as *early succession* and are dominated by rapidly growing plants that do best with full sun and bare soil, and they reproduce quickly. Over time, these plants give way to more shade-tolerant, *late successional* species—typically trees and shrubs—that dominate the site until the next round of disturbance resets the time clock.

The disturbance cycle for vacant urban land tends to be episodic: structures are built, used, abandoned, torn down, and rebuilt. Such disturbance is often tied to governmental approval processes that can take anywhere from 5 to 20 years for implementation (Muratet et al. 2007). For agricultural land, the typical disturbance cycle occurs on an annual basis and begins every time the ground is plowed. Most farming practices are designed to prevent succession from happening in order to allow the growth of annual crops. The disturbance cycle for woodland landscapes is on the order of 50 to 100 years, depending on a combination of unpredictable climatic, biological, and economic factors (Weiss et al. 2005). The periodicity of the disturbance cycle has a major impact on the composition of the spontaneous plant communities that grow in each of the three habitat types: annual and biennial species dominate agricultural landscapes; long-lived woody plants (trees, shrubs, and vines) typically dominate forested landscapes; and the patchwork of annuals, herbaceous perennials, and woody plants found in cities reflects the heterogeneity of its disturbance cycles.

Preadaptation, Ecology, and Evolution

The plants that flourish on abandoned or unmanaged urban land are famous for their ability to grow under extremely harsh conditions. Through a quirk of evolutionary fate, many of them have evolved life history traits in their native habitats that have "preadapted" them to flourish in cities. Indeed, several studies have shown that many common urban plants are native either to limestone cliff habitats and rocky outcrops or to dry, open grasslands with neutral or alkaline soils (Gilbert 1989; Larson et al. 2004; Wittig 2004; Lundholm and Marlin 2006). The authors argue—by analogy—that cities with their tall, granite-faced buildings and concrete foundations are biologically equivalent to the natural limestone cliffs where these species originated. Similarly, they suggest that the increased use of de-icing salts along walkways and highways has produced high pH microhabitats, which are often colonized by species adapted to limestone-rich soils or coastal habitats. In general, preadapted urban plants come from a variety of habitats in nature that typically experience high levels of disturbance, including riverbanks (seasonal flooding), grasslands and prairies (burning or grazing), cliffs and rock outcrops (extreme sun and wind exposure), eroded slopes (unstable soil), and coastal or arid zones (salty soils). Appendix 1 lists some of the habitats in nature that are sources of "preadapted" urban plants.

Preadaptation is a powerful idea for understanding urban plant ecology for two reasons. First, it helps answer questions about why some species are common in cities whereas others are not. Second, it replaces a static, "native" definition of nature based on history and geography with a dynamic, "cosmopolitan" definition based on fitness and flux. In general, plants that can succeed on their own in the urban environment need to be *adaptable* (i.e., non-specialized) in all aspects of their life history from germination through seed production and dispersal; *opportunistic* in their ability to take advantage of resources (mainly water and nutrients) that are available for only a brief period of time; and *tolerant* of the adverse growing conditions created by covering the soil with pavement (Baker 1974; Del Tredici 2010; Knapp et al. 2010). As important as preadaptation is in illuminating the story of urban vegetation, it is really just the introductory chapter in the much longer book of evolution. Modern molecular research has shown that many of our urban plants and animals have undergone measurable changes in their genes (genetic) or in the expression of their genes (epigenetic) after migrating into cities or other drastically disturbed landscapes (Meerts et al. 1998; Cheptou et al. 2008; Donihue and Lambert 2014; Alberti et al. 2017; Johnson et al. 2018; Schilthuizen 2018).

Research has also shown that many of the plants that flourish in urban habitats are hybrids that emerged after their once-isolated parents were brought together by globe-trotting humans. Reports of hybridization among plants that colonize disturbed habitats have become increasingly common as modern transportation technologies have accelerated the breakdown of geographical barriers that once kept species separate (Schierenbeck and Ellstrand 2009; Kelcey and Müller 2011; Johnson and Munshi-South 2017). In general, hybrids seem to be better suited to colonizing disturbed ground than either of their parents and are disproportionally

Preadaptation in action: tree-of-heaven (*Ailanthus altissima*) on the Great Wall of China (*left*) and on a lesser wall in Boston (*right*).

common in landscapes that have been transformed by human activities (Anderson 1952; Thomas 2017).

The history of urban vegetation can be summarized as moving from preadaptation based on evolutionary history, to epigenetic or genetic change through natural selection, and finally to genomic transformation through hybridization. To the surprise of many scientists, cities are providing unprecedented opportunities to study how quickly ecology can be transformed into evolution.

Where Do Urban Plants Come From?

The concept of preadaptation helps answer questions about why some plants are common in the urban environment and others are not, but the question of

Princess tree (*Paulownia tomentosa*) colonizing an abandoned building in New London, Connecticut. From the plants' perspective, a decaying brick wall is just a limestone cliff.

TABLE 1. Geographical Origin of the 268 Species of Plants Described in This Book

	North & Central America	Europe &Central Asia	Eastern Asia	Eurasia & North America	Africa	Totals
Mosses, Ferns and Horsetails	2	—	—	5	—	7 (2.6%)
Conifers	—	—	1	—	—	1 (0.4%)
Woody dicots	32	12	25	—	—	69 (25.7%)
Herbaceous dicots	49	90	6	8	—	153 (57.1%)
Monocots	9	17	2	8	2	38 (14.2%)
Totals	92 (34.3%)	119 (44.4%)	34 (12.7%)	21 (7.8%)	2 (0.7%)	268

how these plants actually reached the city still needs to be answered. Of the 191 *herbaceous angiosperms* (153 dicots and 38 monocots) treated in this book, 57% arrived in North America from Europe (including adjacent parts of Central Asia and North Africa), 30% are native only to North America (including Central America), 8% are native to both North America and Eurasia, 4% originated in Eastern Asia, and 1% from Africa. By contrast, of the 69 *woody angiosperms* covered here, the majority are native to temperate North America (46%) and Asia (36%), with only 17% coming from Europe (Table 1). The differences in the geographical origin of the herbaceous species as compared to the woody species suggest that deep evolutionary patterns underlie the distribution of urban plants (Fridley 2013).

The origin and global dispersal of spontaneous urban vegetation is as much a cultural as a biological phenomenon. European ecologists have traditionally had a stronger interest in urban ecology than their American counterparts because Europe has a much longer history of urbanization. By combining sophisticated archaeological work with modern ecological research, they have been able to reconstruct the complex history of the continent's urban flora. One of the most interesting results of this process is to subdivide the category of alien plants into *archaeophytes* and *neophytes*. The former were introduced with agriculture into a given area from other parts of Europe prior to 1500, whereas the latter were introduced after 1500, mainly from Asia and North and South America (Pyšek 1998; Wittig 2004). This distinction is of little relevance to American botanists, who tend to consider all European plants "aliens."

The dichotomy between archaeophytes and neophytes, however, is important because it reflects the long history—going back thousands of years—of moving

A robust urban forest has developed on this vacant lot in Detroit.

both food and medicinal plants around 1) *within* the Mediterranean basin (between southern Europe, North Africa, Turkey, and the Middle East); 2) from central Asia *into* the Mediterranean basin; and 3) from the Mediterranean basin *into* central, eastern, and northern Europe. When one realizes that at least twenty-two of the plants covered in this book were described almost 2000 years ago by the Greek physician Dioscorides in *De Materia Medica*—a five-volume work about medicinal plants, animals, and minerals that remained in usage into the 1600s—one begins to get an inkling for just how ancient the relationship between archaeophytes and humans is (Appendix 2). The fact that we no longer use these plants to treat diseases does not diminish the significance of the role they played in the development of human culture.

The movement of plants from Europe into North America that began in the 1600s and continued through the 1800s is part of the larger story of human migration. The earliest North American settlers from Europe brought their entire lifestyle along with them—not only their own personal belongings and food for the first year but also seeds of their crop plants, livestock and the fodder to feed them, and medicinal plants. In addition to these intentionally cultivated plants, the colonists inadvertently brought weed seeds embedded in the hay they brought for their animals and mixed in with the grains they sowed on the land they cleared. Essentially, the weeds came to North America along with the crops. In his classic book, *New England's Rarities Discovered* (1672), John Josselyn documented the presence of dozens of European weeds growing spontaneously in New England less than fifty years after the pilgrims first landed under the category, "Of such plants as have sprung up since the English planted and kept cattle in New England" (Appendix 3).

The most extreme manifestation of Europe's biological invasion of North America resulted from the imbalance of trade between the two continents during the nineteenth century when European ships coming to the United States typically arrived filled with rocks and soil for ballast. This material was then discarded on shore before the ship loaded up with cargo for the return trip, creating extensive "ballast

Monet with weeds in Detroit: chicory (*Cichorium intybus*), yellow sweet clover (*Melilotus officinalis*), and spotted knapweed (*Centauria stoebe*).

grounds" in many American port cities. Needless to say, the unique assemblages of plants that showed up on these piles of earth attracted the attention of observant botanists, who carefully documented the arrival and spread of the stowaway species (Mehrhoff 2000).

English-speaking people dominated the initial settlement of eastern North America, but immigrants from other western and eastern European countries were quick to follow, driven by difficult economic or political conditions in their homelands. And just as the members of each group came with their own language and culture, they also brought their own special crops and agricultural practices. Indeed, of the 155 non-native species covered in this book, the vast majority were intentionally cultivated prior to their escape from the farm or garden (Rehder 1946; Mack 2000, 2003; Mack and Erneberg 2002; Starfinger et al. 2003).

This flood of European plants into eastern North America—beginning with the French in Quebec City in 1608, the Pilgrims in Plymouth in 1620, and the Dutch in New Amsterdam in 1625—has permanently altered the landscape. By contrast, relatively few North American species—other than crop plants—have become naturalized in Europe, most notably black locust, black cherry, false indigo, pokeweed, and box elder, along with a number of goldenrod, aster, and evening primrose species, all of which are adapted to disturbed, early successional habitats (Marks 1983; Kowarik 2003; Wittig 2004; Kelcey and Müller 2011). While the asymmetry of the biological exchange between the two continents undoubtedly reflects the lopsided nature of the cultural exchange, it also appears that evolutionary factors related to climate and to a much longer history of agriculture have played a role in giving the European flora an edge in terms of invasive potential (Fridley 2013; Fridley and Sax 2014).

In terms of Asian plants making their way to North America, there was some limited importation of plants directly from China in the 1850s, but many of them first arrived on the East Coast by way of European (mainly English) nurseries. Beginning in 1861, some seven years after Admiral Perry forced Japan to trade with the United States, new plants from that country began coming directly to North America. It took a while for these introductions to become established in the nursery trade, but by the late 1880s a wide variety of Asian plants were commercially available to the general public, albeit mostly wealthy homeowners who could afford their high prices (Del Tredici 2014, 2017a, 2017b). Over time, the use of exotic Asian plants began filtering into suburban landscapes, which expanded rapidly in the late 1800s and early 1900s as cities were becoming more crowded with immigrants. It is worth noting that many of these early Asian introductions, including kudzu, multiflora rose, Japanese barberry, and various honeysuckles, were widely planted in rural landscapes from the 1930s through the 1970s when controlling erosion—a legacy of the Dust Bowl era—was an economic and ecological priority of federal, state, and local governments.

Data are difficult to come by, but at least two studies document the slow but steady increase in the number of non-native species growing spontaneously in the Northeast over time. One study examined the various editions of *Gray's Manual of Botany* between 1856 and 1950 and found an increase in the number of non-native species from 10.7 to 19.9% (Moore et al. 2010). A second study looked at the frequency of herbarium specimens collected over the past hundred years in the greater New York City area and found dramatic increases in the distribution of a number of non-native woody plants as well as concurrent decreases in many native species (Clemants and Moore 2005; Aronson et al. 2015). These studies are significant because they establish a baseline for tracking changes in the composition of vegetation over time and illuminate the increasingly significant role that non-native species are playing as urbanization expands throughout the Northeast.

At the same time that exotic plants were being introduced from Europe and Asia, native American species were being shuffled around as settlers moved inland from the Atlantic Coast. An early example of such translocation is provided by black locust (*Robinia pseudoacacia*), which originally had a distribution limited to the central Appalachian and Ozark mountain ranges and now grows spontaneously throughout most of North America. Beginning in the mid-1700s, farmers outside its native range began planting black locust— a root-suckering species—to provide a sustainable supply of rot-resistant wood for fence posts and timbers. In a cruel

A mature stand of black locust (*Robinia pseudoacacia*) has been enhanced through the process of *intaglio*, or the creation of a landscape by the judicious removal of unwanted plants.

An urban Krakatoa: This sea of urban blacktop is like a volcanic lava flow, and the plants that grow here, including mullein (*Verbascum thapsus*), chicory (*Cichorium intybus*), New England hawkweed (*Hieracium saubadum*), and white heath aster (*Symphyotrichum pilosum*), can tolerate extreme heat and drought.

twist of fate, several states, including Massachusetts, now classify this still desirable timber tree as an invasive species.

In 1979, Viktor Mühlenbach published a monograph on the "synantropic" flora of St. Louis that documented the significant role that railroads played in moving Midwestern prairie plants from west to east during the nineteenth and twentieth centuries. He also showed how plants native to the East Coast as well as those introduced from Europe moved in the opposite direction. This work is significant because it establishes baseline ranges for some native species that are the North American equivalent of European archaeophytes.

These eighteenth- and nineteenth-century manifestations of globalization pale in comparison with the ecological translocations that have taken place during the twentieth century, as formerly independent economies have been connected by various modes of high-speed transportation. People and commercial goods—including many plant and animal products—now flow seamlessly around the globe accompanied by a host of associated weeds, pests, and pathogens (Kareiva et al. 2007). Modern examples of such unintentional exchange include the Asian longhorned beetle, which entered North America in the 1990s embedded in wooden shipping pallets and now threatens maple trees growing in several metropolitan areas in the

Northeast, and the emerald ash borer from China that was first detected in Detroit in 2002 and is now killing ash trees up and down the East Coast. Recent research by forest pathologists have shown that unintentional importation of various pests and pathogens associated with international trade is likely to continue well into the twenty-first century (Roy et al. 2014).

In much the same way that globalization has destabilized long-standing local economic institutions, it has also destabilized long-standing ecological associations and led to the worldwide establishment of what ecologists call *novel ecosystems*. This term was developed to describe the unique communities of plants and animals that have emerged in areas where large-scale human activities—including agriculture, forestry, mining, and infrastructure development—have irrevocably altered as much as a third of the earth's terrestrial ecosystems (Hobbs, Higgs, and Hall 2013). More recently, the novel ecosystem concept has been extended to include areas affected by urbanization (Kowarik 2011; Donihue and Lambert 2014; Thomas 2017). In essence, the plants that grow spontaneously in cities are a cosmopolitan array of species that reflect a tumultuous past and offer a preview of an unpredictable future.

Urban Ecology

At first glance, the term *urban ecology* might seem like an oxymoron. Nevertheless, cities do have a distinctive ecology dominated by the needs of people and driven by socioeconomic rather than ecological factors. People welcome other organisms into cities to the extent that they contribute to making the environment a more attractive, more livable, or more profitable place to be; and they vilify as weeds those organisms that flourish without their approval or assistance. The complex cultural, geological, and biological histories of American cities have turned them into mosaics of ecological habitats, each of which is inhabited by a distinctive suite of plants and animals. Indeed, an enormous variety of non-human life has managed to crowd into cities to form a cosmopolitan collection of organisms that is typically more diverse than that of the surrounding non-urban areas (Knapp et al. 2010; Kelcey and Müller 2011; Müller et al. 2013).

From a strictly functional perspective, urban land can be classified into three broad categories: *remnant natural* landscapes, managed *horticultural* landscapes, and *abandoned ruderal*[1] landscapes (Whitney 1985; Zipperer et al. 1997; Del Tredici 2010) (Table 2). Remnant natural landscapes are left over from the time before the city spread out to embrace them. They include everything from freshwater wetlands and woodlands to salt marshes near the coast. They are residual parts of the original landscape dominated by native species growing on relatively undisturbed soils. Managed horticultural landscapes are created specifically for human use and enjoyment. They are dominated by horticultural plants growing on relatively good

1. The word *ruderal* comes from the Latin word *rudus* (plural *rudera*) meaning broken stone or rubble. In a botanical context, it describes a plant that colonizes disturbed soil, waste ground, or construction debris.

TABLE 2. The Taxonomy of Urban Landscapes

	Remnant natural landscapes	Managed horticultural landscapes	Abandoned ruderal landscapes
Land-use category	Minimally disturbed woodlands, wetlands, and coastal habitats	Large and small parks; cemeteries; lawns; ball fields; residential gardens; corporate, commercial, and educational landscapes	Abandoned industrial land, vacant lots, infrastructure edges, railroad and river corridors, degraded wetlands, abandoned parks, and successional woodlands
Primary vegetation	Native plants and invasive species	Cultivated plants and weeds	Spontaneous native and non-native species
Soil characteristics	Native soils with minimally disturbed profiles; free draining and nutrient poor; high biological diversity and mineral cycling	Nutrient rich and highly manipulated soils; often manufactured or relocated from agricultural areas; moderate biological diversity	Disturbed and/or compacted fill; nutrient poor; often contains construction rubble; may or may not be contaminated; low to moderate biological diversity
Maintenance requirements	Low to moderate	Moderate to intensive	None to low

soil and require a consistent input of human energy (i.e., maintenance) to survive. Abandoned ruderal landscapes receive no maintenance and are dominated by spontaneous vegetation growing mainly on compacted or fill soils. These neglected "wastelands" typically experience high levels of disturbance and can be as inconspicuous as a sidewalk crack or as prominent as an abandoned rail line.

Given that cities have been totally transformed by human action undercuts the notion that they have a native flora that can be restored to its original condition. In the first place, most urban land has been completely altered from what it once was—the native vegetation has been wiped out and the soil structure destroyed. Second, the climate conditions the original flora was adapted to no longer exist. And finally, most urban habitats, such as those constructed on filled wetlands, are strictly human creations with no natural analogs and no indigenous flora. Historically speaking, a native flora once grew where the city now stands, but the idea that this vegetation can be restored to the same site as it exists today is both an ecological and evolutionary impossibility (Gould 1998).

Certainly we can plant native species in the city and they will grow—but only if we provide them with the appropriate soil and maintain them the way we would any other intentionally cultivated plant. In the absence of ongoing horticultural

Post-industrial succession in an abandoned parking lot in Detroit—Is this the latest version of ecology's old field succession?

maintenance (planting, weeding, mowing, and watering), spontaneous vegetation will eventually dominate most urban and suburban landscape plantings (Del Tredici 2007). In general, the amount of spontaneous vegetation one finds in a given city is inversely proportional to its maintenance budget and, by extension, its economic prosperity—Detroit is clearly at one end of the spectrum and Manhattan at the other.

Most people are surprised when they learn that cities—at least when it comes to plants—are more biodiverse than the surrounding countryside. The unexpectedly high species richness found in cities is the product of a number of factors, including habitat heterogeneity, altered climatic conditions, horticultural activity by humans, variation in vertical and horizontal infrastructure, and the opportunities that disturbance provides for the establishment of new species (Zerbe et al. 2003; Sukopp 2004; Kelcey and Müller 2011). A number of European researchers have gone so far as to propose that certain types of inner-city landscapes should be actively conserved because of the key role they play in promoting and maintaining animal biodiversity (Maurer et al. 2000; Kowarik and Körner 2005; Muratet et al. 2007; Müller et al. 2013).

The disturbance-driven cycle of succession in the city is unpredictable and as responsive to changing socioeconomic conditions as to environmental factors. To borrow a term from ecology, the urban environment is *patchier* than either agricultural or forested habitats (Pickett et al. 2008). As anyone who lives in the city knows, construction is a constant: Old buildings are being razed, new buildings are being erected, infrastructure is being repaired, roadways are being repaved, and, most destructive of all, vacant land is being cleared for development. At any given time a significant portion of the urban fabric is in the process of being torn up and rebuilt.

This periodic, unpredictable disturbance combined with the continual introduction of new species into the urban environment from a variety of outside sources—including nursery plants (with their associated weeds); lawn-seed mixes; topsoil from other locations; and seeds carried by wind, people, and migrating

Figure 1. The drivers of the succession cycle in urban environments (with apologies to J. P. Grime 2001).

animals—provide all the components necessary to drive the urban succession cycle (Zipperer 2011). Disturbance and competition, along with ongoing habitat stress, interact on a continuous basis in the urban environment to create a constantly shifting array of plant associations dominated by early successional species (Figure 1). By way of analogy, this situation is not all that different from the dynamism that characterizes the human population of the city as one ethnic group replaces another when the economic status of a given neighborhood shifts upward or downward.

The concept of disturbance, as used in ecology, typically refers to periodic, unpredictable calamities, such as flooding, wildfires, drought, hurricanes, insect infestations, logging, and ice storms. Their results are catastrophic in the immediate time frame, but the ecosystems typically make a rapid recovery toward some semblance of their former state. In contrast to episodic disturbances are the so-called *environmental stress factors* that permanently alter baseline environmental conditions. These are exemplified by such widespread phenomena as climate change, acid precipitation, competition from non-native species, and urbanization. Unlike periodic disturbances, environmental stress factors often lead to the development of novel ecosystems. Native ecosystems are well adapted to recovering from classic natural disturbances but not so good at dealing with long-term, human-caused environmental stresses. Novel ecosystems, by contrast, can handle both ecological disturbance and environmental stress (Hobbs, Higgs, and Hall 2013).

As many scientists have pointed out, modern climate change can be viewed as a massive, uncontrolled experiment on the impact of increased atmospheric carbon dioxide concentrations on the earth's ecosystem. Most people now realize that after nearly 200 years of burning fossil fuels, these impacts have been both wide ranging—they have affected every corner of the globe—and, at the local level, unpredictable. And this is where the cities come in. Because all the impervious paving and buildings absorb and retain heat, and all the cars, air conditioners, and electrical equipment generate heat, the annual mean temperatures of large urban areas (i.e., with populations in excess of a million people) can be up to 5.4° F (3° C) warmer than the surrounding non-urban areas; on extreme occasions the temperature differences between the city and the countryside—particularly at night—can be as high as 21.6° F (12° C). This "heat island effect" means that the core areas of

Chain-link fences can act as "safe sites" for the establishment of plants as exemplified by tree-of-heaven (*Ailanthus altissima*) root sprouts in Boston.

many of our larger cities have already warmed up to the levels predicted for the surrounding countryside 20 to 30 years from now. Indeed, urban areas are the perfect place to study how climate change will affect future environments because they are already experiencing readily observable impacts. In addition, cities are experiencing increases in both growing season length and growing season temperature as well as winters that are shorter and warmer, all of which directly affect plant growth (Sieghardt et al. 2005; Pickett et al. 2011).

While increased temperature is probably the most ecologically significant factor that distinguishes the city from the surrounding countryside, several other features of the atmosphere associated with urban areas can also have profound impacts (both negative and positive) on the growth of plants, including higher levels of air pollution, higher concentrations of carbon dioxide, altered solar radiation regimens, altered wind patterns, and decreased humidity (Gregg et al. 2003; Sukopp 2004; George et al. 2007; Pickett et al. 2011).

Urban soils typically contain higher levels of available nitrogen because of increased levels of atmospheric deposition associated with the burning of fossil fuels (Decina et al. 2017). In addition, the high percentage of land covered with impervious surfaces in urban areas increases stormwater runoff and soil compaction and decreases water percolation, which when combined, exacerbate drought conditions for plants. And finally, the increased use of road salt in northern cities tends to elevate soil pHs and decrease cation exchange capacity and water availability, all of which typically have negative impacts on plant growth (Sukopp 2004; Byrne 2007; Godefroid et al. 2007; Marcotullio 2011).

Recent research on large urban areas in Europe—across gradients of both space and time—have found that up to 40% of the plants that grow in these environments are non-native and that urbanization, in general, favors species that are tolerant of soils that are relatively fertile, dry, sunny, and alkaline (Pyšek 1998; Godefroid et al. 2007; Muratet et al. 2007; Knapp et al. 2010). Clearly the vegetation of the city is a reflection of its distinct environmental conditions.

Ecosystem Services and Ecological Functionality

The evidence is overwhelming that novel ecosystems can restore ecological functionality to habitats that have been drastically disturbed by humans and that they are expanding across the globe as climate change impacts become more pronounced (Walther et al. 2009; Davis et al. 2011; Ellis et al. 2013; Thomas 2017). Nevertheless, debates still rage within horticultural and conservation circles about the value of novel ecosystems, the threats posed by invasive species, and the merits of planting native versus non-native species (Tallamy 2007; Carroll 2011; Murcia et al. 2014). Despite the negative publicity that many of the plants described in this book receive—much of it deserved—research clearly shows that spontaneous vegetation can provide ecological and social services that help make cities more livable for its inhabitants (Pataki et al. 2011; Burkholder 2012; Robinson and Lundholm 2012; Thompson 2015; Riley et al. 2018). For example:

1. Carbon sequestration and oxygen production: All plants perform these functions, but spontaneous vegetation takes care of itself and is therefore more sustainable than cultivated species that require ongoing maintenance.
2. Biodiversity enhancement: Plants provide food and habitat for animals of all kinds; from the ecological perspective, the more flora, the more fauna.
3. Organic matter production: Microbes break down plant litter into its component mineral nutrients that cycle through multiple levels of the soil ecosystem—an especially important function on sites with degraded soil.
4. Erosion control on slopes: Many urban plants require bare soil for germination, an attribute that allows them to quickly colonize and stabilize disturbed ground.
5. Stream and riverbank stabilization: Water levels in most urban (and non-urban) waterways are constantly in flux and the woody vegetation helps keep their banks intact.
6. Stormwater infiltration: Wherever vegetation is growing in soil, water infiltration is better than with pavement or bare, compacted soil.
7. Mitigation of soil pollution: Post-industrial land and urban wetlands are often contaminated with heavy metals and petroleum products; the plants that grow spontaneously on such sites have the capacity to help clean up that environment.
8. Temperature reduction: By creating shade and transpiring water vapor, all trees significantly lower air temperatures at the local scale, thereby making cities healthier and more livable for its inhabitants.
9. Social and educational contributions: "Urban wilds" provide people with unfettered access to nature and create opportunities for social interaction not typically allowed in managed parklands.

Although the value of most of these ecological services is self-evident, a few of them deserve additional comment. The term *phytoremediation* refers to the ability of plants and their associated microbes to help clean up contaminated sites by

A stand of Norway maples (*Acer platanoides*) dominates a typical highway slope in Watertown, Massachusetts.

selectively absorbing and sequestering high concentrations of heavy metals, such as cadmium, lead, copper, zinc, chromium, and nickel, in their tissues and degrading petroleum products (Porębska and Ostrowska 1999). Several common urban species—most notably prickly lettuce (*Lactuca serriola*), lambsquarters (*Chenopodium album*), and mugwort (*Artemisia vulgaris*)—possess this ability to an unusual degree and, unbeknownst to virtually everyone, are helping to "detoxify" polluted, vacant land at no cost to the taxpayer. In wetland habitats, common reed (*Phragmites australis*) is one of the most-effective plants for removing dissolved nitrogen and phosphorous from polluted wetlands, and it is used for tertiary wastewater treatment in many third-world countries; it also sequesters a variety of heavy metals.

As regards urban biodiversity, an exhaustive review of the literature on the vegetation of fifty-four Central European cities concluded that "large cities are characterized by a higher species-richness in terms of vascular plants than the surrounding rural areas and the species-richness is increasing with the population size of the city" (Kelcey and Müller 2011, 582). They also determined that these cities contained between 20% and 60% non-native species, with a mean value of 40.3%, and that this figure was 13.7% higher than the ratio of non-native to native plants in the surrounding countryside. Other vegetation surveys of European cities indicate that the ratio of non-native to native species as well as the ratio of neophytes to archaeophytes increase as one gets closer to the more highly disturbed parts of the city (Pyšek 1998; Pyšek et al. 2004). The clear trend is that with increasing urbanization, the world is becoming less native and more cosmopolitan (Wittig and Becker 2010; Müller et al. 2013). In urban environments, the traditional dichotomy between native and non-native species has lost much of its relevance as well as its power to explain what is actually happening to the world.

Recent research has demonstrated that trees—both planted and spontaneous—

Two "pollarded" American elms (*Ulmus americana*) have adapted to this chain-link fence niche in Hartford, Connecticut.

can significantly improve urban air and water quality and, by extension, people's overall health and well-being (Nature Conservancy 2016). Other studies have shown that interacting with plants, even if it is just looking out a window at them, can improve people's physical and psychological health (Barton and Pretty 2010). In low-income neighborhoods, which rarely include maintained public parks other than ball fields, access to areas of unmanaged vegetation provides opportunities for interacting with nature as well as social interaction (Keil 2005; Jorgensen and Keenan 2012).

On the negative side, spontaneous vegetation can be seen as a threat to public health because it provides habitat for a number of animals, such as rats, mosquitoes, and ticks, that serve as vectors for pathogens and infectious diseases (Garvin et al. 2012; Gulachensik et al. 2016). Under the influence of climate change, the health impacts of some toxic plants are projected to get worse. Controlled experiments with two native species—ragweed (*Ambrosia artemisiifolia*) and poison ivy (*Toxicodendron radicans*)—have shown that elevated levels of carbon dioxide induce the former to produce significantly more of its highly allergenic pollen and cause the latter to produce higher concentrations of its rash-producing toxins (Ziska et al. 2003; Mohan et al. 2006). While these results do not bode well for human health in a carbon dioxide–rich future, they do provide an important reminder of the innate capacity of "weeds" to take advantage of the mess that people have made of the planet (Ziska and Dukes 2011).

In a related vein, the illicit activities that often take place in minimally managed urban wilds—including drinking, doing drugs, having sex, painting graffiti, and camping out—often produce complaints from other users of the sites who feel threatened by these activities as well as by the fear of attackers hiding in tall vegetation. To the extent that urban landscapes dominated by spontaneous vegetation are perceived as threatening, they fit within the modern concept of a "wilderness" that is defined as any tract of land that lies outside the boundaries of human control (Hofmeister 2009; Del Tredici and Rueb 2017; Kowarik 2018).

It is a foregone conclusion that our environment will continue to deteriorate over the next few decades as people continue to pump more heat-trapping carbon dioxide into the atmosphere and more acid rain falls back to earth to pollute the water and the soil. The worldwide migration of people from the countryside into cities is also contributing to environmental degradation because land that was once covered with vegetation is being covered by buildings and pavement that respectively generate and retain heat. The confluence of climate change and urbanization—acting in concert with the global spread of non-native species—has set the stage for spontaneous vegetation to play a major role in reshaping the urban as well as the rural landscapes of the future (Walther et al. 2009).

Living with Spontaneous Urban Vegetation

Most urban residents tend to interpret the presence of spontaneous urban vegetation in their neighborhood as a manifestation of neglect. Yet the identical plants growing in a suburban or rural context are typically described as "wildflowers"—just think about the beautiful combination of chicory and Queen Anne's lace growing along a country road in July and August. As per the earlier discussion about weeds, the context in which a plant grows has everything to do with how people feel about it. The same species that people despise as invasive in North America are often cherished natives in their homeland (Coates 2006). The common reed (*Phragmites australis*) is widely distributed in river deltas throughout Europe and serves as a major bulwark against coastal erosion. The fact that it is dying throughout much of its native range is viewed as an ecological crisis, whereas the same species growing in the same type of habitat in eastern North America is perceived as a noxious weed that should be eradicated (Saltonstall 2002).

This relativity is explained by the fact that people are looking at the same plant through the subjective lens of an ecological value judgment that places a higher value on the nativity of a given plant than on its function. This dichotomy may be appropriate and necessary for preserving wilderness areas or the habitats of rare species, but it does not work well in an urban context, where cultural, economic, and ecological functionality should be recognized as having a greater value than the restoration of historical ecosystems (Rink 2005; Sagoff 2005).

How city dwellers respond to the presence of spontaneous vegetation in their midst is influenced by both personal preferences and cultural norms. In many European cities, residents' feelings about this vegetation is divided—some welcome it as a manifestation of unrestrained nature while others see it as an indicator of dereliction that should be removed. Such responses have led to the categorization of urban dwellers as either "nature lovers" or "neat freaks." Interestingly, the percentage of people in each camp can vary from one city to the next within the same country (Keil 2005; Rink 2005; Weber et al. 2014). In general, Europeans have a

In August, Queen Anne's lace (*Daucus carota*) and chicory (*Cichorium intybus*) make a stunning combination along roads in the Northeast.

greater tolerance of and appreciation for spontaneous urban vegetation in their cities than Americans do, as reflected in their use of the term "urban nature" to describe it (Kowarik 2018). In his fascinating book *Darwin Comes to Town* (2018), Menno Schilthuizen goes so far as to suggest that letting spontaneous vegetation grow freely promotes urban ecology and in the long term, urban evolution.

Nonetheless, spontaneous urban vegetation often leaves something to be desired from the aesthetic point of view, and most people perceive it as ugly or messy and an indicator of neglect (Nassauer and Raskin 2014; Riley et al. 2018). This fact raises the question of whether a way could be found to harmonize a recognition of the functionality of spontaneous vegetation with people's desire to live in a neat and tidy environment. As is the case with urban ecology in general, the Europeans have been hard at work on this question for the past thirty years. *The Dynamic Landscape* (Dunnett and Hitchmough 2004) introduces North American readers to years of European work on ways to manipulate the aesthetic characteristics of spontaneous vegetation through the judicious addition or deletion of species. The book is unique in combining solid ecological research on natural grasslands and agricultural meadows with the horticultural goal of creating aesthetically pleasing, low-maintenance landscapes. Indeed, landscapes with spontaneous vegetation fit the technical definition of *sustainable* in the sense that they are adapted to the site, require minimal maintenance, and are ecologically functional (Kühn 2006). More recently, Peterken (2013) has produced the definitive book on the cultural and natural history of European meadows that lays down a solid ecological foundation for people interested in adapting this ancient landscape form to modern urban conditions.

Using works such as these as a guide, the concept of a "cosmopolitan urban meadow" presents a way of capitalizing on the aesthetic and ecological opportunities presented by some of the herbaceous species covered in this book (Appendix 4).

Creative preservation of spontaneous vegetation: European silver birch (*Betula pendula*) growing amid abandoned railroad tracts in Berlin's Natur-Park Südegelände (*left*); a large eastern cottonwood (*Populus deltoides*) is the only spontaneous tree remaining after the development of Brooklyn Bridge Park in New York City (*right*).

The basic idea behind the meadow is to put together an assemblage of plants that will 1) grow well on typical urban soil, 2) create aesthetically pleasing combinations when in bloom, and 3) hold the ground against competitors until such time as an alternative use for the land emerges (Del Tredici and Luegering 2014). Clearly, some minimal soil preparation is required to get the meadow established, but certainly much less than it would take to start a lawn from seed. Some selective weeding during the early stages of growth is necessary to keep unsightly competitors, such as ragweed and mugwort, from dominating. And finally, an annual mowing in late summer or fall is required to keep woody plants from taking over.

The aesthetics of landscapes dominated by spontaneous *woody plants*, with their longer life spans and less conspicuous ornamental attributes, are more difficult to manipulate than those with *herbaceous species*, which can be mowed down when all else fails. Few people realize that the appearances of an urban woodland can be greatly enhanced by selectively removing trees or shrubs that are unsightly, in poor condition, or too close together. Such landscape "editing," if done by a trained horticulturist, can dramatically improve both the appearance and the recreation potential of a spontaneous urban woodland (Kowarik and Langer 2005). The foundation for such an approach to vegetation management was laid out in 1966 by Frank Egler in *The Wild Gardener in the Wild Landscape,* a remarkable book he wrote under the

TABLE 3. **Managing Novel Ecosystems through Deletions and Insertions**

Common *Intaglio* **Deletions** for novel ecosystems should include:

– High-climbing vines that can strangle trees (e.g., wild grapes, bittersweet, wisteria)
– Diseased, seriously damaged, and/or hazard trees
– Plants that are unfriendly or unhealthy for people (e.g., poison ivy, multiflora rose, Japanese barberry, ragweed)
– Unsightly and/or aggressive plants that are commonly perceived as indicators of neglect (e.g., mugwort, *Ailanthus*, Japanese knotweed, *Phragmites*)

Sustainable plant **Insertions** for novel ecosystems should be:

– Tolerant of prevailing site conditions
– Require minimal applications of pesticides, herbicides, or fertilizers in order to look good
– Able to flourish without supplemental water
– Non-invasive, that is, they will not spread aggressively into the surrounding natural areas

pseudonym Warren G. Kenfield. Egler was light-years ahead of the present generation of "ecological restorationists" in advocating the use of herbicides to remove unwanted trees in order to arrest or suspend forest succession at a particular stage of succession. He called this process *intaglio* after the engraving technique that creates a positive image by removing unwanted material from the surface of a blank plate. It is a process of design by removal as opposed to design by insertion (Table 3).

Another landscape concept that relies heavily on the use of spontaneous vegetation was introduced in the landmark book *Redesigning the American Lawn* (Bormann, Balmori, and Geballe 1993, 57). The *freedom lawn*, as the authors define it, is a collection of low-growing, spontaneous plants that "results from an interaction of naturally occurring processes and the selective effects of lawn mowing." The book goes on to contrast this diverse plant community with the *industrial lawn*, which requires a substantial inputs of chemical fertilizers, weed killers, and water in order to maintain a monoculture of Kentucky bluegrass. The beauty of the freedom lawn concept is that it transforms a wasteful sink of petroleum products into a sustainable source of ecological benefits. Indeed, the power of spontaneous vegetation to make urban wasteland productive is nothing short of miraculous (Figure 2).

Criteria for Including Plants in This Book

Developing the list of plants covered in this book was a lengthy process that required me to develop a biologically relevant definition of the term *urban*. From a plant's perspective, it is the abundance of paving rather than the density of the human population that defines the urban environment (Arnold and Gibbons 1996). In

Mirabile Dictu
(Wonderful to Tell: Virgil)

Figure 2. Undated PUNCH cartoon by H. M. Bateman.

other words, a roadside edge is a roadside edge whether it is in a city or a suburb. Recent research has shown that when the level of impervious surface in a given area exceeds 25 to 30%, it is urbanized from an ecological perspective (Raciti et al. 2012).

For the purposes of this book, the term *urban* does not include minimally disturbed, usually protected, natural areas found within the boundaries of the city, and the species list does not include native plants that are confined to such remnant habitats, although it does include species that have moved out of these areas and into other parts of the city. Nor does the species list include cultivated plants characteristic of managed landscapes within the urban environment unless they have shown some capacity to "escape" and become naturalized in surrounding unmanaged areas.

The 268 plants covered in this book are commonly found growing spontaneously in the cities of the Northeast—both large and small—from Montreal, Quebec, in the north, to Boston, Massachusetts, in the east, to Washington, D.C., in the south, and to Detroit, Michigan, in the west (Figure 3). The core coverage area

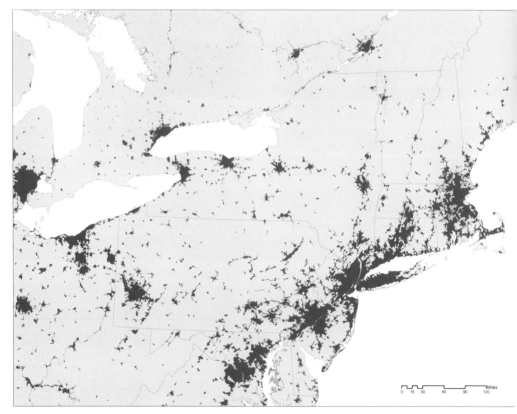

Figure 3. Range map for this book showing the urban areas of northeastern North America. This map is based on census blocks with a population density of at least 1,000 people per square mile and surrounding census blocks with an overall density of at least 500 people per square mile. U.S. data are from the 2000 census; Canada data are primarily from the mid-1990s (Government of Canada). Illustration courtesy of Alan Berger and Case Brown of the Project for Reclamation Excellence (P-REX).

extends along the eastern seaboard from Boston to Philadelphia. My categorization of a plant as common is based on a review of the regional literature on spontaneous urban vegetation, local floras, and my personal observations over nearly twenty years in numerous northeastern cities.

The decision to exclude certain plants was ultimately a subjective one on my part, motivated partly by space limitations and partly by the book's focus on species that are common in the urban environment. As one moves south and west of the core coverage area, one is more likely to find spontaneous species not covered in this book. From the ecological perspective, most of the plants included in this book are disturbance-adapted and early successional (Appendix 5). Excluded are many familiar ground cover species, such as the common orange daylily, lily-of-the-valley, pachysandra, vinca, and goutweed, all of which can spread aggressively when planted in gardens but are typically not capable of crossing large special gaps without being physically transported by people. In other words, they are aggressive but not invasive species. In this second edition of the book, I have added a number of plants that become more common within the region since 2010. I have also added a number of species that should have been included in first edition but for whatever reasons were not. In total, this book covers 20% more species than its predecessor.

How to Use This Book

Wild Urban Plants of the Northeast was written for urban residents who have some curiosity about the plants that grow in the city—the type of person who admires tenacious weeds growing in the sidewalk cracks or trees growing out of building foundations. The extensive use of photographs coupled with a minimal use of botanical jargon is intended to facilitate plant identification for people without a formal botanical background. The book takes the position that every plant that grows in the city is interesting and can tell us something about the site conditions where it is growing as well as about its complex connections to human history. If one knows how to read the signs—a process that begins by learning the names of the plants—spontaneous vegetation becomes an open book on urban nature and a curious commentary on human nature.

Following the format of *Weeds of the Northeast* by Uva, Neal, and DiTomaso (1997), I have arranged the plants in this book according to major botanical categories. The highest level is the Taxonomic Group, which includes: *mosses, ferns, horsetails, conifers, woody dicots, herbaceous dicots*, and *monocots*. Within each of these groups, the plants are arranged alphabetically by family, and within each family individual species are listed alphabetically. Appendix 6 describes the key characteristics of twelve common plant families—out of a total of sixty-nine—that account for 57% of the species covered in this book. Being able to recognize the higher-order characteristics of a given plant family is a critical step not only in determining the identity of a specific species but also in getting a handle on the daunting diversity of the plant kingdom.

While determining the correct SCIENTIFIC NAME of a plant may seem to be a straightforward process to the non-botanist, it is anything but. Modern molecular technology has led to a revolution in plant taxonomy, which has produced new scientific names for a number of species whose nomenclature had been unchanged for a hundred years or more. In the ten years between the two editions of this book, a significant number of species names have had to be changed along with a number of Family designations. For accuracy's sake, it is also important to give the name of the person who first proposed the name—the author of the name if you will. In this regard, it is worth noting that plant names followed by the capitol letter "L." were named by Linneaus, whose 1753 book, *Species Plantarum*, marks the official start of modern binomial nomenclature.

To check for the latest updates on ever-changing SCIENTIFIC NAMES, I used

The Plant List (http://www.theplantlist.org/) managed by an international consortium of botanical gardens and conservation organizations as well as the Go Botany website managed by the New England Wildflower Society (https://gobotany.new englandwild.org/). The online Flora of North America (http://www.efloras.org/flora_page.aspx?flora_id=1) was a great resource for accurate plant descriptions, taxonomy, and distribution information. And finally, The Biota of North America Project (BONAP) website (http://bonap.net/napa) provided distribution maps for many of the species treated in this book. When there was a lack of agreement as to the correct name of a plant among various websites, I made an executive decision about which one to use.

The species' accounts also list a few of the most important SYNONYMS for both scientific and common names that have been used for the plant over the past hundred years or so. To list them all would require way too much space. Equally problematic to determining the correct scientific name is deciding on a plant's COMMON NAME, which typically varies by region as well as by time period. To help resolve this potentially confusing situation, I have relied heavily, but not exclusively, on the standardized common names provided on the Plants Database of the U.S. Department of Agriculture (http://plants.usda.gov/).

The PLACE OF ORIGIN of a given species is not always easy to pin down, especially for plants with a long history of association with humans. For this information, I have relied on a number of different sources, the most important of which was the *Manual of Vascular Plants of Northeastern United States and Adjacent Canada* (1991) by Gleason and Cronquist. In addition, a number of internet websites, such as Wikipedia, provided current information on the origin of some of the plants listed in this book.

I have tried to keep the technical descriptions of the plants' morphological characteristics simple and jargon-free, a goal that is possible only because of the many photographs that accompany the text. In those instances where jargon was unavoidable, the GLOSSARY at the end of the book defines all the technical terms in the text. I used separate categories for VEGETATIVE CHARACTERISTICS and FLOWERS AND FRUIT in order to make it easier to access information. In this regard, several books were critically important. For trees, I relied heavily on Barnes and Wagner's *Michigan Trees* (2004) and for herbaceous plants, I used Uva, Neal, and DiTomaso's *Weeds of the Northeast* (1997) and *Weeds of Lawn and Garden* by John Fogg Jr. (1945). I particularly admire the latter book not only for its concise treatment of species and accessible text but also for the beautiful and informative line drawings by Léonie Hagerty. A number of outstanding websites cover plant identification issues in detail, most notably the Illinois Wildflowers website: http://www.illinoiswildflowers.info/.

Information on GERMINATION AND REGENERATION is critical to understanding the ecology of any plant. Reproduction from seed or from some vegetative part of the plant can provide important life history traits that are seldom covered in field guides but are essential to understanding how a species becomes established and persists in stressful ecological conditions.

Under the subhead HABITAT PREFERENCES, I explain under what conditions the plant prefers to grow and where it can be found both in natural as well as urban habitats. The category ECOLOGICAL FUNCTIONS describes the ecological services provided by the plant in the urban environment and is included in most of the woody plant treatments and in about half of the herbaceous species. Information for both of these categories came from the author's experience as well as a wide variety of books, articles, and internet sources. The category CULTURAL SIGNIFICANCE describes a plant's multifaceted relationship with humans—mainly medicinal, food, and utilitarian uses—along with information about how and when the plant first arrived in North America. For many people, especially non-scientists, this is the most interesting part of the plant treatments.

The category RELATED SPECIES describes common urban plants in the same botanical family as the primary entry. Although usually there is just one related species per entry, I did include as many as three in a few cases because they are all abundant in the urban environment. And finally, in a few of the entries, I have included the category SIMILAR SPECIES to indicate another plant—usually a native species not described in the book—with which it might be confused.

In conclusion, it should be noted that I have not included information on how to kill any of the described plants. Every other book on weeds and/or invasive species, to say nothing of the internet, provides more than enough information on this topic. Indeed, the authors of these works seem to assume that the main reason for learning to identify weeds is to be able to kill them. My decision not to discuss eradication is in keeping with this book's two primary goals: to teach people how to identify the plants that are growing spontaneously in urban areas and to counter the widespread perception that these plants are ecologically harmful and should be eliminated from the landscape. This is not to say that I think these species should be actively planted, rather that people should appreciate them for what they are contributing to areas of the city that no one is taking care of—simply put: if it's green, it's good.

A Note on the Photographs

Except where noted in the Acknowledgements, all photographs in the book were taken outdoors by the author with a digital camera (either an Olympus 8080, a Canon Digital Rebel XTi, a Canon EOS Rebel T6i with a 17–85 mm macro lens—or an Apple iPhone 6 in a pinch). Photos were selected for their ability to capture the essence of the plant growing in its natural habitat, with an emphasis on important macroscopic features that are useful in identifying a plant at various stages in its life cycle. The extensive use of photographs has allowed the author to keep written descriptions to a minimum and allows a novice naturalist confronted with an unknown plant to ignore the formal classification system and simply "go fishing" for a match.

Bryum argenteum Hedw.   Silvertip Moss

SYNONYMS: silvery bryum, silver moss, silvergreen bryum moss

LIFE FORM: **perennial moss**; mat-forming ground cover

PLACE OF ORIGIN: Silvertip moss is a truly cosmopolitan species that is found throughout the Northern and Southern Hemispheres, including Antarctica. It is one of the most widespread and distinctive mosses in the world.

VEGETATIVE CHARACTERISTICS: Silvertip moss produces a dense mat of gray-green leaves that can grow to a height of approximately 0.5 inch (1.2 cm) tall with a spread of about 2 inches (5 cm). Although mosses can be difficult to identify, this species is easy to recognize because of the silvery appearance of its foliage, which is caused by light reflecting off the translucent leaf tips. Like all mosses, this species absorbs moisture through its foliage and lacks true roots.

REPRODUCTION: Silvertip moss mats are mostly sterile, but occasionally they produce pendant red spore capsules that release *haploid* spores that grow into the typical moss mat. Separate male and female silvertip moss plants produce haploid sperm and egg cells (the *gametophytes*), respectively. In the presence of water, the sperms "swim" a small distance and one of them fuses with an egg cell to form a *diploid* zygote at the top of the female gametophyte. As it develops, the diploid zygote (the *sporophyte*) is carried above the mat of foliage in the capsule where it undergoes *meiosis* to produce thousands of tiny haploid spores which, following dispersal, restart the life cycle.

GERMINATION AND REGENERATION: In addition to haploid spores, silvertip moss produces bulbils (*buds*) in the axils of its leaves that disperse readily and form new mats identical to their parent.

HABITAT PREFERENCES: Silvertip moss is a common moss in urban environments, particularly in sunny, dry locations. It grows on a variety of alkaline substrates with little or no soil, including the edges of decaying blacktop where sand collects, between brick pavers, and on weathered, flat roofs.

RELATED SPECIES: **Silky bryum (*Gemmabryum* [*Bryum*] *caespiticium* (Hedw.) J.R. Spence)** is a common urban species whose emerald green foliage has a distinct sheen and is tightly compacted into dense clumps; it rarely produces spore capsules. Silky bryum can be somewhat weedy and typically grows in dry, alkaline microsites, such as sidewalk cracks and the spaces between paving stones. It reproduces vegetatively from *gemmae*, *bulbils*, and *tubers*, all of which facilitate its spread. Depending on the time of year, **purple moss or fire moss (*Ceratadon purpureus* (Hedw.) Brid.)** forms a dull green, yellow-green, or reddish-brown mat of foliage that expands slowly over time. It grows in sunny, disturbed sites with acidic soil and has a special affinity for sites that have been burned. In urban areas, purple moss is found on decaying asphalt, weathered roofs, and compacted soil. In non-urban areas, it grows on rocks, sand dunes near the ocean, and mine tailings contaminated with heavy metals.

Silvertip moss on its typical gravelly substrate

Spore capsules of silvertip moss

Silky bryum has shiny, emerald-green foliage

Silky bryum in its typical urban habitat between paving blocks

Purple moss in early April, between the curb and an asphalt sidewalk

Purple moss in cross section with spore capsules

Onoclea sensibilis L. Sensitive Fern

SYNONYM: bead fern

LIFE FORM: **herbaceous perennial**; up to 2 feet (60 cm) tall

PLACE OF ORIGIN: eastern North America

VEGETATIVE CHARACTERISTICS: Tall, leathery leaves (*fronds*) are roughly triangular in outline and can reach up to 2 feet (60 cm) long. They are coarsely dissected into 12 or more opposite pairs of leaflets and those at the base are more lobed than those near the apex. Young, expanding leaves are pale red; mature leaves are light green with scattered white hairs on the underside. The foliage dies with the first frost (hence the common name). A forking, brown rhizome, which grows just below the surface, gives rise to new shoots and the fibrous root system.

REPRODUCTIVE STRUCTURES: Sensitive fern reproduces from spores that are produced on stiff fertile fronds that stand approximately 1 foot (30 cm) tall. These fronds have numerous short branches that are lined with small, bead-like capsules that release the spores. The fertile fronds are green during the growing season and turn dark brown at maturity; they typically remain standing through the winter.

GERMINATION AND REGENERATION: The haploid spores germinate on moist soil and develop into tiny gametophytes that produce egg and sperm cells which, following fertilization, give rise to the diploid sporophyte generation that we recognize as a fern. Once established, sensitive fern spreads by means of branching rhizomes, eventually forming large, long-lived clumps.

HABITAT PREFERENCES: Sensitive fern is more sun tolerant than other ferns and reproduces freely under urban conditions. It grows best in sunny, moist sites but can also be found in dry, shady situations. It is common along the margins of freshwater wetlands, ponds, streams, and swamps; drainage ditches; woodland understories; urban meadows; and railroad rights-of-way.

RELATED SPECIES: **Hayscented fern (*Dennstaedtia punctilobula* (Michx.) Moore) (Family: Dennstaedtiaceae)**, a North American native, is very different in appearance from sensitive fern but is also common in the urban environment, especially in dry, shady woodlands. Its leaves are about 16 inches (40 cm) long and are divided into 20 or so pairs of thin leaflets that are in turn divided into subleaflets that bear small "spore dots" on their undersides. The bases of the fronds are dark brown or black and lack papery scales. Under favorable conditions, hayscented fern can spread rapidly from rhizomes and form large clumps. It sprouts back rapidly following disturbance and dies back to a perennial rootstock with the first frost. The plant is remarkable for its ability to colonize shady sites with very little soil. Because of its drought-tolerance and unpalatability to deer, hayscented fern seems to be increasing in abundance in forest understories where it grows.

Emerging leaves of sensitive fern growing at
the edge of a pond

Sensitive fern frond

Dry fertile fronds
of sensitive fern

Dense stand of sensitive fern

Hayscented fern growing in a moist
microclimate

Hayscented fern fronds

Equisetum arvense L. Field Horsetail

SYNONYMS: common horsetail, horsetail fern, bottle brush, jointed rush

LIFE FORM: **herbaceous perennial**; up to 1.5 feet (45 cm) tall

PLACE OF ORIGIN: Europe, Asia, and North America

VEGETATIVE CHARACTERISTICS: Stiff, hollow, green stems emerge in late May; as the season progresses, they produce whorls of secondary branches that give them a "bottle-brush" appearance. Scale-like true leaves, which are located at the base of the secondary branches, are inconspicuous. The stems are rough to the touch and die down to the ground in late fall.

REPRODUCTIVE STRUCTURES: In early spring, field horsetail will occasionally produce unbranched fertile stems up to 1 foot (30 cm) tall before the foliage emerges. These stems—the diploid sporophytes—are brownish to pale pink and are topped with a small cone, the *strobilus*, which releases thousands of haploid spores before withering away. Horsetails are considered the most primitive of the fern-type plants that produce spores rather than seeds.

GERMINATION AND REGENERATION: Field horsetail spores germinate to produce a haploid gametophyte that produces eggs and sperm that unite and give rise to the diploid horsetail stem. Established plants can form large clumps that increase in size by producing new stems from deep growing rhizomes.

HABITAT PREFERENCES: This species is commonly found in sandy soil along roadsides as well as in areas with poor drainage. It can form large clumps along highway banks covered with coarse gravel and among the ballast supporting railroad tracks because it is resistant to many of the herbicides used to kill flowering plants.

CULTURAL SIGNIFICANCE: Native Americans ate the young shoots of field horsetail and used a tea made from the stems to treat kidney and bladder disorders. European traditional medicine used an extract of the stems as a diuretic and an astringent. Field horsetail accumulates silica in its tissues—up to 12% of its dry weight—and also accumulates various heavy metals, including selenium, gold, and mercury, which suggests that it may be useful for cleaning up contaminated soil (phytoremediation). Extinct horsetail species were abundant during the Carboniferous period (250 million years ago) and much of the coal we burn today is composed of their compressed remains.

RELATED SPECIES: **Scouring rush** (*Equisetum hyemale* L.) is native to Europe, Asia, and North America and consists entirely of green stems, lacking the whorled secondary branches of field horsetail. While scouring rush is common along railroad tracks (because of its resistance to herbicides), it is typically found in shady, moist areas. The plant sequesters silica from the soil in its stems, which gives them an abrasive quality that makes them ideal for scrubbing pots and pans on camping trips; hence its common name. Scouring rush is a popular landscape plant in southern California where its tight, vertical growth habit compliments the severe modern architecture.

Field horsetail
vegetative stem

Field horsetail growing on a gravel-covered highway embankment

Field horsetail spore-
producing stems

Cultivated scouring rush in Los Angeles

Scouring rush growing in railroad ballast

Scouring rush spore-
producing cones

Taxus cuspidata Sieb. & Zucc.   Japanese Yew

SYNONYM: spreading yew

LIFE FORM: evergreen tree; 10–50 feet (3–15 m) tall

PLACE OF ORIGIN: northeast Asia

VEGETATIVE CHARACTERISTICS: Cultivated Japanese yews, grown from cuttings, have a spreading, shrubby form with dense foliage; spontaneous trees that grow from seed tend to develop a single trunk with widely spaced branches and sparse foliage. Dark, evergreen needles are arranged alternately along the stem, are short stalked, and are 0.5–1 inch (1.2–2.5 cm) long. Its distinctive reddish-brown bark peels off in thin strips.

REPRODUCTIVE STRUCTURES: All yews are dioecious, with separate male and female individuals. On both sexes, the cones develop in late fall in the axils of the needles and go through the winter in a rudimentary state of development. Male cones are produced in abundance and are clearly visible in fall along the underside of the year-old shoots. Female plants produce far fewer cones than do male plants. Male cones shed their pollen in late spring, and female plants produce bright red "fruits"—about 0.25 inch (8 mm) long—in the fall, which birds readily consume and disperse. The fruit consists of a hard, brown seed (which is highly toxic) surrounded by a fleshy red aril (which is nontoxic).

GERMINATION AND REGENERATION: Seeds germinate in spring; seedlings grow best in light to dense shade and soils rich in humus.

HABITAT PREFERENCES: Since the 1980s, Japanese yews have started to appear with increasing frequency in the shady understory of disturbed urban forests and edges of landscape plantings. Seedlings grow slowly and can persist in a suppressed condition for many years. It is one of the few conifers that flourishes without cultivation in the urban environment.

CULTURAL SIGNIFICANCE: Most Japanese yews growing spontaneously in the urban environment have arisen from seed of the cultivar 'Nana', which was introduced as an ornamental into North America from Japan in 1861 by Dr. George Rogers Hall of Bristol, Rhode Island (Del Tredici 2017a). This cultivar went on to become a ubiquitous hedge and foundation plant in the Northeast because of its low, spreading growth habit and because it is hardier than the English yew (*T. baccata*). In the early 1980s, the anti-ovarian cancer drug taxol was isolated from the bark of the Pacific yew (*T. brevifolia*) and subsequent research has determined that the compound occurs in the foliage of all *Taxus* species.

Japanese yew growing spontaneously in the understory of a Norway maple woodland

Japanese yew foliage

Japanese yew seedling in a typical urban habitat

Well-pruned planting of Japanese yews in the Boston area

Male cones of Japanese yew along the undersides of 1-year-old shoots

Mature female cones of Japanese yew

Rhus typhina L.   Staghorn Sumac

SYNONYMS: *Rhus hirta*, velvet sumac, vinegar tree

LIFE FORM: **multistemmed shrub**; up to 30 feet (9 m) tall in the wild

PLACE OF ORIGIN: eastern North America

VEGETATIVE CHARACTERISTICS: Staghorn sumac produces sparsely branched, crooked stems and 1-year-old twigs that are densely covered with velvety hairs (hence the common name). Alternate, pinnately compound leaves are composed of 5–15 pairs of leaflets with serrated margins and are 1–2 feet (30–60 cm) long. In October the leaves turn spectacular shades of red, orange, and yellow. In urban environments, staghorn sumac typically grows 10–20 feet (3–6 m) tall.

FLOWERS AND FRUIT: Staghorn sumac produces terminal panicles of greenish-white, insect-pollinated flowers in late spring on separate male and female plants. Cone-shaped clusters of deep red, velvety fruits, up to 6 inches (15 cm) long, terminate the branches of female plants in late summer and persist through the winter.

GERMINATION AND REGENERATION: Seeds are consumed and dispersed by numerous species of birds and germinate in sunny, open sites; established plants produce root suckers and form large, dense thickets.

HABITAT PREFERENCES: Staghorn sumac grows best in full sun and well-drained soil. In the urban environment, it is common in minimally maintained public parks, vacant lots, rubble dumps, degraded or emergent woodlands, rock outcrops and stone walls, unmowed highway banks, and railroad rights-of-way.

ECOLOGICAL FUNCTIONS: Tolerant of roadway salt and compacted soil; food and habitat for wildlife; erosion control on slopes.

CULTURAL SIGNIFICANCE: Juice from the crushed berries, when steeped in water, can be used as a gargle for sore throat or as a refreshing, somewhat acidic, summer tonic (the hairs on the surface of the seeds contain malic acid). Native Americans used the leaves and fruits as a poultice to soothe irritated skin. Staghorn sumac is widely planted for erosion-control purposes.

RELATED SPECIES: **Smooth sumac (*Rhus glabra* L.)** is somewhat smaller than staghorn sumac (up to 15 feet [5 m] tall) and lacks velvety hairs on its purplish stems and petioles. It sprouts vigorously from the roots, forming large clumps, and its alternate leaves turn brilliant red in the fall. Darlington and Thurber (1859, 79) noted that "this shrub is apt to be abundant in neglected sterile old fields; and its prevalence, in arable lands, is strong evidence of the occupant being a poor thriftless farmer." Native Americans and European immigrants used extracts of the bark, leaves, and fruit to treat asthma and a variety of other ailments, as well as for tanning leather and making dyes.

Staghorn sumac foliage and fruits

New staghorn sumac stems develop from root sprouts and can form large colonies

Staghorn sumac fruits ready for dispersal

Distinctive hairy stems of staghorn sumac

Smooth sumac's outstanding fall color

Hairless, light purple, smooth sumac stems

Toxicodendron radicans (L.) Kuntze   Poison Ivy

SYNONYMS: *Rhus radicans, Rhus toxicodendron,* poison vine

LIFE FORM: **deciduous woody vine**; up to 50 feet (15 m) long

PLACE OF ORIGIN: eastern North America

VEGETATIVE CHARACTERISTICS: This ubiquitous and highly variable vine can climb tall trees; grow as a ground cover; or form a dense, spreading shrub. Regardless of its growth habit, all of its branches have a distinct horizontal orientation. The older climbing stems produce conspicuous aerial roots that give them a "bearded" appearance (the Latin name *radicans* means rooted stems). Alternate, compound leaves are composed of 3 glossy leaflets (source of the old adage, "leaves of three, let them be"), each approximately 4 inches (10 cm) long with smooth or coarsely toothed margins; the terminal leaflet always has a petiole whereas the lateral leaflets are nearly sessile. Leaves of plants growing in the shade turn dull yellow in fall; plants in full sun turn bright red.

FLOWERS AND FRUIT: Poison ivy produces clusters of small 5-petaled, yellowish-green, insect-pollinated flowers in the axils of the leaves from May through June on separate male and female plants. The berries on female plants turn from green to gray to white when they mature in September.

GERMINATION AND REGENERATION: Fruits are eaten by birds, and seeds germinate beneath their roosts. Established plants spread by underground rhizomes and by stems that root where they touch the ground.

HABITAT PREFERENCES: Poison ivy grows best in moist soils in shade or full sun but can also be found in dry, rocky or sandy sites. Its tolerance of high salt concentrations accounts for its abundance near the ocean as well as along busy roadsides. In the urban environment, it is common on rock outcrops and stone walls; climbing up telephone poles, buildings, and chain-link fences; along unmowed highway banks and railroad tracks; and climbing up tree trunks in moist or dry woodlands.

ECOLOGICAL FUNCTIONS: Tolerant of roadway salt and compacted soil; food and habitat for wildlife; erosion control on slopes.

CULTURAL SIGNIFICANCE: Touching any part of this plant, in summer or winter, causes allergenic dermatitis in 60–80% of people. The offending ingredient is *urushiol*, which is located in the sap. Once absorbed through the skin it causes a characteristic itchy rash within a day or two of contact. Poison ivy has been used in traditional medicine and was one of the herbs sold by the Shakers to treat chronic paralysis, rheumatism, and skin diseases. In 1624, Captain John Smith became the first European to describe the plant, which "being but touched causeth rednesse, itching, and lastly blisters, the which howsoever after a while passe away of themselves without further harm" (Hauser 1996, 27).

Aerial roots give poison ivy stems a distinctive "bearded" appearance

Poison ivy foliage

Poison ivy will climb on anything

The mature, horizontal growth habit of poison ivy growing on an iron fence

Poison ivy fruits ripen in September

Poison ivy in full fall color along a roadside chain-link fence

Berberis thunbergii DC   Japanese Barberry

SYNONYM: Thunberg's barberry

LIFE FORM: **deciduous shrub**; 2–6 feet (0.6–2 m) tall

PLACE OF ORIGIN: Japan

VEGETATIVE CHARACTERISTICS: Stems of Japanese barberry are grooved and brown; the alternate leaves are approximately 0.5 inch (1.2 cm) long and spoon- or egg-shaped with smooth edges. An unbranched, sharp spine at each node makes the whole plant potentially painful to touch. The twigs, when broken, are bright yellow, as are the roots and rhizomes. Japanese barberry is one of the first shrubs to leaf out in early spring and one of the last to lose leaves in the fall when they turn various shades of yellow, orange, or scarlet.

FLOWERS AND FRUIT: Small, yellow, cup-shaped flowers have 6 petals and are produced in April and May; they line the length of the stem, hanging down from leaf axils in small clusters; and they are pollinated by insects. The showy red fruits, which are about 0.5 inch (1 cm) long, mature in the fall and persist into early winter. Plants growing in the sun produce much more fruit than do those in the shade.

GERMINATION AND REGENERATION: Seeds are dispersed by birds and germinate in sunny or shady conditions. Stems root where they touch the ground and established plants spread by underground rhizomes.

HABITAT PREFERENCES: Japanese barberry is a highly adaptable shrub that grows well in dry soil and full sun as well as in moist shade. Because of this adaptability, it has been widely planted in low-maintenance landscapes—such as parking lot islands and commercial properties—where it persists long after most other plants have died. In the woodland understory, it conspicuously leafs out in spring before native species and retains its foliage in the fall longer than the surrounding plants. Deer generally avoid eating Japanese barberry.

CULTURAL SIGNIFICANCE: This Japanese species was introduced into North America as an ornamental in 1875 in the form of seeds from the St. Petersburg Botanical Garden in Russia to the Arnold Arboretum in Boston. The first documented reports of the plant spreading on its own are from 1910. Cultivars with colored foliage, especially purple ('Crimson Pygmy') and yellow, are still widely planted. The fruit can be used to make jelly, and the bright yellow roots, which contain the medicinally active compound *berberine*, have been used as a digestive tonic and a purgative. Many states list Japanese barberry as an invasive species because of its ability to spread aggressively in the forest understory, but historical research has shown that in forests where the plant is now dominant, it was probably established many decades earlier when the land was open pasture (DeGasperis and Motzkin 2007).

Japanese barberry leafs out in spring, well before most native woody plants

Japanese barberry flowers and foliage

Bright yellow roots of Japanese barberry are rich in the medicinally active compound *berberine*

Japanese barberry in the forest understory produces fall color in early November

Japanese barberry produces fruits in the autumn

Alnus glutinosa (L.) Gaertner   Black Alder

Synonyms: *Alnus vulgaris*, European alder

Life Form: deciduous tree; 35–60 feet (10–18 m) tall

Place of Origin: Europe, western Asia, and North Africa

Vegetative Characteristics: Black alder produces shiny, alternate leaves that are round or oval with wavy margins, prominent veins, and a small notch at the tip. They are 2–4 inches (5–10 cm) long and nearly as wide, and they typically stay green late into the autumn, when they fall without changing color. Bark of the mature tree is dark brown or black with flaking plates and vertical fissures.

Flowers and Fruit: Separate male and female flowers are produced on the same plant (*monoecious*). The pendulous male catkins are conspicuous in early spring when they elongate and shed their pollen; the woody female "cones" persist on the trees through winter and spring and help to identify this species.

Germination and Regeneration: Seeds are dispersed by water or wind and germinate in the spring on moist, bare ground in full sun; established trees sprout readily from the base following damage. In Europe, trees are typically cut down to the ground on a 10-year cycle (coppice forestry) to produce poles and small-diameter fuel wood.

Habitat Preferences: Black alder is common along stream and riverbanks and can form dense thickets in freshwater wetlands.

Ecological Functions: Soil improvement; tolerant of road salt and saturated soil; food and habitat for wildlife; stream and riverbank stabilization.

Cultural Significance: Black alder was originally introduced for ornamental, forestry, and soil stabilization purposes. It also improves soil quality by fixing atmospheric nitrogen in symbiosis with an actinobacteria in the genus *Frankia* in its roots. In Europe, the bark and the young shoots have long been used as dyes, and the leaves for tanning leather. It also had limited use as a medicinal plant to treat swelling and inflammation, and in the past its rot-resistant wood has been used to make underwater pilings that support buildings.

Similar Species: Three species of alder are native to the northeast—***Alnus incana**, **A. serrulata**,* and ***A. viridis**.* They grow in moist to wet habitats and are typically shrubby and multistemmed as opposed to single-trunked trees. All of these species fix atmospheric nitrogen and none are common in the urban environment.

Male (*pendulous*) and female (*upright*) black alder flowers

Black alder keeps its foliage longer than most native trees

Male flowers of black alder are conspicuous in very early spring

Black alder leaf

Spent female cones (fruits) of black alder from the previous year (*left*) along with the current season's cones (*right*)

Betula nigra L.   River Birch

SYNONYM: red birch

LIFE FORM: **deciduous tree**; 30–60 feet (10–18 m) tall

PLACE OF ORIGIN: eastern North America

VEGETATIVE CHARACTERISTICS: When young, river birch produces a peeling, buff or light tan colored bark; as the tree ages, the bark becomes dark reddish-brown or black with deep furrows, and it breaks off in irregular plates. Under natural conditions, the tree typically develops a multistemmed growth habit. Alternate, simple leaves—up to 3.5 inches (9 cm) long by 2.5 inches (6 cm) wide—are roughly diamond- to egg-shaped, sharp pointed, and have serrated margins. The leaves turn yellow in the fall.

FLOWERS AND FRUIT: River birch produces separate male and female flowers on the same plant (monoecious); the pendulous male catkins are conspicuous in early spring when they expand to a length of 2–3 inches (5–7.5 cm) and shed their pollen to the wind; the female catkins mature quickly, releasing their seeds in late spring.

GERMINATION AND REGENERATION: Wind- and water-dispersed seeds germinate immediately on bare ground that is exposed after high spring water levels recede; saplings and mature trees sprout readily from the base.

HABITAT PREFERENCES: River birch grows best in moist soil with full sun and prefers the edges of freshwater wetlands and pond, stream, and riverbanks. It is one of the most heat tolerant of all birch species, a trait that predisposes it to grow well in urban environments.

ECOLOGICAL FUNCTIONS: Heat reduction in paved areas; tolerant of roadway salt and compacted soil; food and habitat for wildlife; stream and riverbank stabilization.

CULTURAL SIGNIFICANCE: This species has become an extremely important ornamental plant in urban areas over the past 30 years, mainly because of its heat tolerance and its resistance to insect pests, including the bronze birch borer and leaf miners that typically damage or kill white-barked birches. The cultivar 'Heritage' is widely planted because it maintains its attractive, buff-colored bark for a few more years than is typical for the species.

Mature river birch along the Charles River in Watertown, Massachusetts

River birch male (*pendulous*) and female (*upright*) flowers

River birch foliage

Distinctive bark of a river birch sapling

River birch stabilizing a riverbank

Betula populifolia Marshall   Gray Birch

SYNONYMS: poverty birch, poplar birch, old field birch

LIFE FORM: **deciduous tree**; up to 35 feet (10 m) tall

PLACE OF ORIGIN: eastern North America

VEGETATIVE CHARACTERISTICS: Gray birch typically develops a multistemmed growth habit, with trunks leaning in various directions over time. On young trees, the bark is a shiny, reddish-brown; on mature trees, it is chalky white, fairly tight (i.e., not peeling), and marked with prominent dark, triangular patches below the branches. Alternate leaves are 2–3 inches (5–7.5 cm) long; they are more or less triangular in shape, tapering to a long point, with coarsely toothed margins. They are glossy, dark green in summer and turn yellow in the fall. Individual gray birch stems are short-lived, typically dying when they reach a diameter of 6–8 inches (15–20 cm). The wood of old trunks lying on the ground decays faster than the rot-resistant bark that is often left intact as a hollow cylinder.

FLOWERS AND FRUIT: Male catkins are produced at the ends of the branches and shed their wind-dispersed pollen in early spring; female catkins produce cylindrical "cones" of seeds, 0.75–1.25 inches (1.8–3 cm) long, which disintegrate over the course of the fall and winter.

GERMINATION AND REGENERATION: Wind-dispersed seeds germinate in a variety of disturbed, sunny sites; mature trees typically sprout from the base to form a cluster of crooked or leaning stems, often bent low by the weight of snow.

HABITAT PREFERENCES: Gray birch is a pioneer species that is tolerant of poor soil and drought. It is found growing primarily on compacted or sandy, acid soil, often along roadsides and old fields in association with quaking aspen. In the urban environment, it is common on rock outcrops, stone walls, railroad rights-of-way, and the edges of rivers and streams.

ECOLOGICAL FUNCTIONS: Gray birch is tolerant of contaminated soil and is known to sequester heavy metals in its foliage and wood (*phytoremediation*). It provides food and habitat for wildlife; erosion control on slopes; and stream and riverbank stabilization.

RELATED SPECIES: **European silver** or **white birch (*Betula pendula* Roth)** is native to Europe and Asia and typically grows to 40–50 feet (12–15 m) tall with a similar growth habit but more pendulous branches than the gray birch has. It was widely planted as an ornamental in the early to mid-20th century but has fallen out of favor because of its susceptibility to pests and pathogens. It grows spontaneously in urban areas along the edges of disturbed woodlands, rivers, and roadways. Like gray birch, it is tolerant of soils contaminated with heavy metals, and it readily colonizes coal mine tailings in Europe.

Male (*pendulous*) and female (*upright*) flowers of gray birch

Mature, multistemmed growth form of gray birch

Gray birch foliage

Gray birch sapling

Stand of gray birch along the Charles River in Watertown, Massachusetts

Trunk of European silver birch in Berlin, Germany

Catalpa speciosa (Warder) Warder ex Engelm.   Northern Catalpa

SYNONYMS: Indian bean tree, cigar tree, western catalpa

LIFE FORM: **deciduous tree**; up to 80 feet (25 m) tall

PLACE OF ORIGIN: central North America

VEGETATIVE CHARACTERISTICS: Northern catalpa has an upright growth habit with a strong central leader that is carried well up into a relatively narrow crown. It is coarsely branched with stout twigs, and the brownish bark has deep fissures separated by flaky ridges. Large, heart-shaped leaves taper to a sharp "drip-tip" and can be up to 14 inches (35 cm) long. Leaves are mostly opposite but can occur in whorls of 3 toward the tips of the stems; they are a distinctive light green during the growing season and turn pale yellow in the fall. The winter twigs display distinctive large, round leaf scars and prominent lenticels.

FLOWERS AND FRUIT: Northern catalpa produces upright, terminal panicles of showy white flowers in late May or June. The tubular blossoms, which are approximately 2 inches (5 cm) wide, have spreading lobes with wavy margins and are marked with 2 yellow stripes and numerous purple spots that serve as "nectar guides" for pollinating bees and moths. A mature, roadside tree in full bloom is a spectacular sight. The slender fruits are more than a foot (30 cm) long but only 0.5 inch (1.2 cm) wide; they are produced in abundance and persist on the tree through the winter, providing an unmistakable clue for identification. The pods contain numerous papery, thin seeds with a fringe of hairs at each end.

GERMINATION AND REGENERATION: Wind-dispersed seeds germinate on bare soil in sunny locations; saplings sprout readily from the base following traumatic injury.

HABITAT PREFERENCES: In its native habitat, northern catalpa is tolerant of alkaline soils and grows best in moist, sunny locations, including floodplains. In an urban context, it occupies exposed sites including roadways, chain-link fences, foundation cracks, and riverbanks; along highways it is a common colonizer of exposed rock cuts.

ECOLOGICAL FUNCTIONS: Heat reduction in paved areas; tolerant of roadway salt and compacted soil; erosion control on slopes; food and habitat for wildlife.

CULTURAL SIGNIFICANCE: This midwestern native was first described as a distinct species in 1853 and became widely popular in the East and Midwest in the late 1800s and the early 1900s because of its cold hardiness, beautiful flowers, rapid growth, and rot-resistant wood that was useful for fence posts (Del Tredici 1986). The tree fell out of fashion when people began to criticize it for being messy and coarse. Native Americans used a tea made from the bark as an antiseptic wash, a laxative, a sedative, and an antidote to snakebite; they also made poultices from the leaves to treat wounds and bruises.

RELATED SPECIES: **Common catalpa (*Catalpa bignonioides* Walt.)** is native to the southeastern United States and is a smaller, less upright, and less hardy tree than northern catalpa. In the Northeast, it does not spread spontaneously like northern catalpa does.

Developing fruits of northern catalpa

A northern catalpa sapling growing on the grounds of an abandoned factory in Connecticut

Northern catalpa flowers

Northern catalpa foliage

Opposite leaf arrangement of this northern catalpa seedling creates a pagoda-like effect

Fruits of northern catalpa remain on the tree through the winter

Celtis occidentalis L.　Hackberry

SYNONYMS: raisin tree, sugarberry, northern hackberry, nettle tree, beaver wood

LIFE FORM: deciduous tree; up to 60 feet (15 m) tall

PLACE OF ORIGIN: eastern North America

VEGETATIVE CHARACTERISTICS: Hackberry is a small- to medium-sized tree with a short trunk and an irregular crown; its gray-brown bark is distinctively covered with corky warts or ridges. Leaves are alternate, simple, 2–4 inches (5–10 cm) long by 1.5–2 inches (3.7–5 cm) wide, and rough to the touch. Leaves are finely serrated, taper to a point at the tip, are lopsided (unequal) at the base, and turn light yellow in fall. Dense clusters of twigs known as "witches' brooms" typically develop on the upper branches.

FLOWERS AND FRUIT: Hackberry produces inconspicuous, wind-pollinated flowers in spring as the leaves are unfolding. Solitary female flowers are produced on slender stalks in the axils of the leaves; they are followed by small, dark purple fruits approximately 0.25 inch (6 mm) in diameter, each containing a single large seed. Fruits ripen in the fall and can persist on the tree through the winter.

GERMINATION AND REGENERATION: Hackberry fruits are consumed and dispersed by birds, and the seeds germinate on a variety of sites. Established trees sprout readily from the stump, and roots will produce shoots following traumatic injury to the main stem.

HABITAT PREFERENCES: Hackberry is a highly adaptable species that can grow in a wide variety of habitats and exposures. In its native habitat, it prefers moist bottomlands with good light and limestone outcrops; in the urban environment, hackberry grows in heavy or gravelly soils with either a high or a low pH. Although not particularly common, hackberry can be found growing along the margins of streams and rivers, roadsides, rock outcrops and stone walls, and railroad tracks. Once established it can persist for many years.

ECOLOGICAL FUNCTIONS: Tolerant of roadway salt and compacted soil; food and habitat for wildlife; erosion control on slopes; heat reduction in paved areas.

CULTURAL SIGNIFICANCE: Hackberry fruits are edible in the fall and are said to taste like raisins or dates. While seldom planted on the East Coast, hackberry is commonly cultivated in the Midwest because of its tolerance of heavy, high pH soils.

Hackberry foliage

Hackberry can be identified in winter by the distinctive "witches' brooms" in its crown

Hackberry growing in Central Park, New York City

Hackberry's characteristic gray, warty bark

Hackberry fruits

Lonicera japonica Thunb.   Japanese Honeysuckle

SYNONYMS: *Nintooa japonica*, Hall's honeysuckle

LIFE FORM: semi-evergreen vine (liana); up to 30 feet (9 m) long

PLACE OF ORIGIN: temperate eastern Asia

VEGETATIVE CHARACTERISTICS: Japanese honeysuckle climbs up and over other plants, smothering them with its luxuriant growth. Stems are reddish-brown and hairy when young; older stems have flaky bark that peels off in strips. Ovate, hairy leaves are opposite and 1–3 inches (2.5–7.5 cm) long with entire margins. Plants are evergreen in the South and deciduous in the North.

FLOWERS AND FRUIT: Japanese honeysuckle produces tubular flowers in pairs on short stalks from May through September, depending on their latitude; the highly fragrant flowers are white at first but change to dull yellow as they age, producing a dramatic bicolor effect. The nectar-rich flowers are pollinated by bees and moths. Its black fruits are approximately 0.25 inch (6 mm) wide and mature from late summer into fall.

GERMINATION AND REGENERATION: Fruits are readily consumed and dispersed by birds, and seeds germinate under a wide variety of conditions from sun to shade and moist to dry. Runners and trailing stems root where they touch the ground, forming large clumps over time.

HABITAT PREFERENCES: Japanese honeysuckle is an early successional species that flourishes in disturbed open areas, floodplain thickets, and woodland edges. In urban sites it is found on chain-link fences, masonry walls, unmowed highway banks, and railroad rights-of-way. The plant grows most vigorously in full sun but is also quite shade tolerant.

ECOLOGICAL FUNCTIONS: Tolerant of roadway salt; food and habitat for wildlife; erosion control on slopes.

CULTURAL SIGNIFICANCE: A tea made from the flowers of Japanese honeysuckle has long been used for medicinal purposes in Asia. The plant was introduced into North America from Japan in 1862 by Dr. George Rogers Hall of Bristol, Rhode Island (Del Tredici 2017a). It quickly became a popular ornamental because of its attractive, sweetly fragrant flowers. Japanese honeysuckle has been spreading on its own since the early 1890s and is now dominant in disturbed woodland habitats throughout the Southeast and, more recently, the Northeast. Many states now list it as an invasive species.

Japanese honeysuckle in bloom in Boston

Japanese honeysuckle flowers

Japanese honeysuckle keeps its foliage in winter

The twining stems of Japanese honey suckle can overwhelm other plants

Japanese honeysuckle is evergreen in central New Jersey

Japanese honeysuckle fruits

Lonicera morrowii A. Gray   Morrow's Honeysuckle

SYNONYM: *Lonicera tatarica* var. *morrowii*

LIFE FORM: **deciduous shrub**; 6–12 feet (1.8–3.6 m) tall

PLACE OF ORIGIN: Japan

VEGETATIVE CHARACTERISTICS: Morrow's honeysuckle is a multistemmed shrub with hollow stems and twigs and pale brown bark that peels off in thin strips. The opposite, dull green leaves are nearly hairless, oblong to egg-shaped, and 1.5–2.5 inches (3.7–7 cm) long. Plants leaf out very early in the spring, well before most native shrubs, and keep their leaves long after most native species have shed theirs in the fall. Morrow's honeysuckle thus has a growing season that is 2–4 weeks longer than most of its competitors.

FLOWERS AND FRUIT: Morrow's honeysuckle flowers abundantly in May and June. The nectar-rich flowers are produced in pairs and are pollinated by a wide variety of insects. They are white when they open but become yellow as they age; they are followed in August by pairs of round red fruits approximately 0.25 inch (6 mm) wide that often rest on top of the leaves. Under full sun conditions, a plant can be totally covered with fruit.

GERMINATION AND REGENERATION: The red berries are readily consumed and dispersed by birds, and seeds will germinate under a wide range of soil and light conditions. Stems root where they touch the ground, creating large, genetically uniform clumps.

HABITAT PREFERENCES: Morrow's honeysuckle grows best in full sun and moist soil but will tolerate drought and shade. It is common along disturbed woodland edges, wetlands, swamps, railroad tracks, rock outcrops, and roadsides.

ECOLOGICAL FUNCTIONS: Tolerant of roadway salt; food and habitat for wildlife; erosion control on slopes.

CULTURAL SIGNIFICANCE: Morrow's honeysuckle is named after the botanist, Dr. James Morrow, on board Commodore Matthew Perry's notorious expedition to Japan in 1854, and seeds of the plant were most likely introduced into North America in 1875 by Thomas Hogg (Del Tredici 2017a). It was widely planted as an ornamental in the 1920s and 1930s for soil and wildlife conservation purposes and subsequently escaped to become a common roadside and forest understory plant. It is now listed as an invasive species in many states.

RELATED SPECIES: **Amur honeysuckle (*Lonicera maackii* (Rupr.) Maxim.)** is a deciduous, multistemmed shrub native to northeast Asia that grows to approximately 15 feet (5 m) tall. Its opposite leaves are 2–3 inches (5–7.5 cm) long and taper to a narrow point, a feature that distinguishes it from Morrrow's honeysuckle. It produces white flowers fading to yellow in July and red berries in the fall that are readily eaten and dispersed by birds. Amur honeysuckle grows well in sun or shade and is common in a variety of disturbed habitats most notably forest edges and understories; it is more common in the Great Lakes region than in the Northeast. **Tatarian honeysuckle (*Lonicera tatarica* L.)** from Central Asia has pink to red flowers and can hybridize with Morrow's honeysuckle.

Morrow's honeysuckle in flower

Morrow's honeysuckle foliage

Flowers of Morrow's honeysuckle

Morrow's honeysuckle fruits

Amur honeysuckle in flower

Amur honeysuckle in fruit

Celastrus orbiculatus Thunb.   Oriental Bittersweet

SYNONYMS: Asiatic bittersweet, round-leaved bittersweet

LIFE FORM: **deciduous vine (liana)**; stems up to 60 feet (18 m) long

PLACE OF ORIGIN: northeast Asia

VEGETATIVE CHARACTERISTICS: While young, the twisted, woody stems of Oriental bittersweet are brown with warty lenticles; they turn gray as they age and develop into stout trunks up to 4 inches (10 cm) in diameter. Most of the alternate, simple leaves are 2–3 inches (5–7.5 cm) long by 1–2 inches (2.5–5 cm) wide, rounded to egg-shaped with a short-pointed tip; but leaves produced by vigorous first-year stems can have long, tapering tips. Oriental bittersweet lacks tendrils and climbs by means of its twining stems, which can eventually strangle its host. Roots are bright orange, and the leaves turn a clear, bright yellow in the fall.

FLOWERS AND FRUIT: This species produces separate male and female individuals (dioecious). Both sexes produce small clusters of inconspicuous, insect-pollinated flowers in the leaf axils in May and June; female plants are covered with round, green fruits that become highly conspicuous in the fall when they turn yellow and split open to reveal orange-red seeds.

GERMINATION AND REGENERATION: Bittersweet seeds are widely dispersed by birds and germinate readily under shady conditions. Stems can produce roots where they touch the ground, and roots can produce numerous shoots (suckers) following damage to the aerial stems. Once established, bittersweet is extremely difficult to eradicate because of its root-suckering ability.

HABITAT PREFERENCES: This highly adaptable vine can grow under a wide variety of light and soil conditions; young plants are extremely shade tolerant. Bittersweet is notorious for its capacity to strangle and overwhelm surrounding trees and shrubs, and it can be highly destructive in woodlands. In urban areas, it is typically found on chain-link fences and stone walls, along unmowed highway banks and railroad tracks, and climbing up the trunks of low-branched trees and shrubs along the edges of minimally maintained woodlands.

ECOLOGICAL FUNCTIONS: Food and habitat for wildlife; erosion control on slopes.

CULTURAL SIGNIFICANCE: Oriental bittersweet's attractive fruits—which are toxic—are widely used for holiday decorations, and the bright orange bark of the root is used in traditional Chinese medicine. The plant was introduced into North America by Thomas Hogg in 1875; it was being widely planted by the 1890s and was reported to have escaped from cultivation by 1912 (Del Tredici 2014). Oriental bittersweet was planted to stabilize highway banks during the 1960s and 1970s. Since then it has spread rapidly throughout the Northeast and is now listed as an invasive species in many states. **American bittersweet (*Celastrus scandens*)** is threatened with extinction due to hybridization (pollen swamping) with Oriental bittersweet.

A botanical boa constrictor—Oriental bittersweet strangling a black locust tree

Oriental bittersweet in fall color in Boston

Oriental bittersweet leaves

A curtain of Oriental bittersweet foliage

Mature fruits of Oriental bittersweet

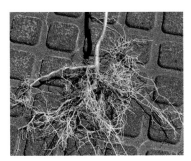

Bright orange roots of Oriental bittersweet

Euonymus alatus (Thunb.) Sieb.   Burning Bush

SYNONYMS: *Euonymus alata*, winged euonymus, winged wahoo, winged staff tree

LIFE FORM: **deciduous shrub**; up to 16 feet (5 m) tall and equally broad

PLACE OF ORIGIN: temperate East Asia

VEGETATIVE CHARACTERISTICS: Burning bush has a wide-spreading, multi-stemmed growth habit; its branches—green turning to brown—produce thin, corky "wings" that can project out from the stems as much as half an inch (1.2 cm) creating a dramatic effect (the Latin word *alatus* means winged). Dark green leaves are opposite, elliptical, 1–3 inches (2.5–7.5 cm) long, and have very short petioles. In fall, the leaves of plants growing in full sun turn brilliant red (hence the common name) whereas those in shade turn rosy pink.

FLOWERS AND FRUIT: Inconspicuous, greenish-yellow, insect-pollinated flowers are produced from May to June. These are followed by reddish-purple, 4-part fruit capsules, which split open in September to reveal 4 dangling orange-red "seeds" (technically the fleshy seed coat is an aril that surrounds the true seed).

GERMINATION AND REGENERATION: Seeds are consumed and dispersed by birds, and they germinate well in shady, understory situations.

HABITAT PREFERENCES: Burning bush is a highly adaptable species that grows well in any combination of sun, shade, dry, or moist conditions and can tolerate a broad range of soil pHs. As a spontaneous plant in the urban environment, it is most common in the understory of disturbed woodlands.

CULTURAL SIGNIFICANCE: Burning bush was introduced into North America as an ornamental plant in the 1870s. It has been widely planted in a variety of low-maintenance landscape situations because of its adaptability and its spectacular fire-engine-red fall color. The cultivar 'Compacta', which typically stays below 6 feet (1.9 m) tall and has smaller "wings" on its branches, is common in ornamental landscapes. Several northeastern states list burning bush as an invasive species.

RELATED SPECIES: **European spindletree (*Euonymus europaeus* L.)** is taller than burning bush and forms either an open, upright shrub or a small tree; its branches lack corky wings. It leafs out early in the spring and holds its leaves late into the autumn when they turn reddish-purple (in the sun) or greenish-yellow (in the shade). In gardens, European spindletree produces a striking autumnal display of pendulous pink capsules with dangling orange seeds. **Wintercreeper (*Euonymus fortunei* (Turcz. Hand.-Maz.)** is a shade-tolerant, evergreen species from eastern Asia that grows either as a climbing vine or as a ground cover. In the Midwest, the plant has escaped cultivation and spread into disturbed forest openings with the help of birds that disperse its seeds; in the Northeast, it is less invasive, but the plant can persist in gardens long after they have been abandoned. Cultivars with yellow and white variegated foliage are common in ornamental landscapes.

Growth habit of a spontaneous burning bush shrub

Foliage and corky wings on the stems of burning bush

Burning bush mall planting in fall color

European spindletree has minimal fall color

Wintercreeper climbing a tree

Elaeagnus umbellata Thunb.   Autumn Olive

SYNONYMS: Japanese oleaster, cardinal olive, Japanese silverberry

LIFE FORM: **deciduous shrub**; to 20 feet (6 m) tall

PLACE OF ORIGIN: temperate East Asia

VEGETATIVE CHARACTERISTICS: Autumn olive typically develops multiple trunks with smooth, gray bark; the smaller branches have stout thorns. Dull green leaves are alternate; 2–4 inches (5–10 cm) long; elliptical to egg-shaped; and have smooth, wavy margins. The stems, the buds, and the undersides of the leaves are densely covered with silvery white to rusty brown scales. In late fall, the leaves turn dull yellow.

FLOWERS AND FRUIT: In May and June, autumn olive produces clusters of fragrant, funnel-shaped, cream-colored to yellow flowers that are approximately 0.3 inch (8 mm) long and pollinated by bees and other insects. Flowers are followed by small, round fruits that are silvery with tiny brown scales at first but turn red at maturity. Fruit is produced in great abundance, often completely covering the stems.

GERMINATION AND REGENERATION: A wide variety of birds and mammals consume and disperse the fruit. Seeds germinate on exposed soil in unmowed grassy areas; established plants sprout vigorously from the base following damage.

HABITAT PREFERENCES: Autumn olive is quite drought tolerant and grows best in full sun and sandy soils. At present, it is more common along roadways in the suburbs and the countryside than in urban areas, but this could change in the future. It is common along sunny forest edges, abandoned pastures, and grasslands near the coast. A symbiotic relationship with bacteria in the genus *Frankia* results in the development of root nodules that "fix" atmospheric nitrogen into ammonia, allowing the plant to grow in nutrient-poor soils.

CULTURAL SIGNIFICANCE: Autumn olive was widely planted along interstate highways in the 1960s and '70s for erosion control, wildlife habitat, and as an ornamental. It has been spread widely by birds, and many states now list it as an invasive species. The edible fruits can be used to make an interesting jelly.

RELATED SPECIES: **Russian olive (*Elaeagnus angustifolia* L.)** is native to western and central Asia and forms a tree that can grow to 30 feet (9 m) tall. Narrow, alternate leaves are dull green above and strikingly silver-gray below. It produces clusters of small, silvery flowers in May or June followed by edible, 0.5 inch (1.2 cm) long, yellow fruits covered with silvery scales. As an ornamental, Russian olive's main attribute is its silver-gray twigs and foliage. Because of its beauty, cold hardiness, drought and wind tolerance, and nitrogen-fixing ability, Russian olive was extensively planted to control erosion and as a windbreak in central and western portions of the U.S. With help from birds that consume and disperse its seeds, Russian olive has escaped from cultivation and is now considered an invasive species in many areas.

Autumn olive taking over in an abandoned baseball field in Vernon, Connecticut

Autumn olive flowers and foliage

Autumn olive growth habit

Spontaneous Russian olive in Detroit

Autumn olive fruits are attractive to birds

Amorpha fruticosa L.   False Indigo Bush

SYNONYMS: indigo bush, bastard indigo, leadplant, leadwort

LIFE FORM: **deciduous shrub**; up to 12 feet (3.6 m) tall

PLACE OF ORIGIN: most of North America, except the Northeast and the Northwest

VEGETATIVE CHARACTERISTICS: False indigo bush is a tall, multistemmed shrub with alternate, compound leaves consisting of 5–15 pairs of gray-green, oval leaflets, each 0.75–1.25 inches (1.8–3 cm) long. Leaves turn pale yellow before they fall.

FLOWERS AND FRUIT: Spikes (racemes) of bee-pollinated flowers are 3–8 inches (7.5–20 cm) long and produced in early summer. Individual maroon-colored flowers are quite small with prominent yellow-orange anthers; these are followed by clusters of small pods—each less than 0.5 inch (1.2 cm) long—that mature in fall.

GERMINATION AND REGENERATION: False indigo bush pods are dispersed by water. Seeds germinate on moist, bare soil; established plants sprout readily from the base to produce dense clusters of tall, flexible stems.

HABITAT PREFERENCES: False indigo bush grows best in full sun, especially along the margins of freshwater marshes, swamps, streams, and rivers. It produces root nodules in symbiotic association with *Rhizobium* bacteria, which allows it to "fix" atmospheric nitrogen and survive on low-nutrient soils.

ECOLOGICAL FUNCTIONS: Stream and riverbank stabilization; erosion control on slopes; salt tolerance; soil improvement; food and habitat for wildlife.

CULTURAL SIGNIFICANCE: Because of its extensive root system, false indigo bush has been widely planted for soil stabilization purposes along riverbanks. Because the plant can be cut down to the ground with impunity when it grows too tall, it is particularly useful in areas where views need to be preserved. It readily escapes from cultivation and is considered an invasive species in parts of Europe and Asia where it was once widely cultivated. Its flexible stems have traditionally been used to make baskets, and a blue dye was extracted from the plant by early European settlers as a substitute for indigo (hence its common name). The flowers attract bees and the plant is considered excellent for honey production. Recent research suggests that false indigo bush produces compounds that may be useful in treating diabetes.

False indigo bush growing along a reinforced shoreline of the Hudson River in New York

Foliage and flowers of false indigo bush

False indigo bush in flower

False indigo bush's muted fall color

Developing seedpods of false indigo bush

Gleditsia triacanthos L.   Honey Locust

SYNONYMS: sweet locust, thorny locust, honeyshucks

LIFE FORM: deciduous tree; 70–90 feet (21–27.5 m) tall

PLACE OF ORIGIN: central North America

VEGETATIVE CHARACTERISTICS: Honey locust can be readily identified by its dark green, once- or twice-compound leaves, which can be up to a foot (30 cm) long; the foliage is finely textured with individual leaflets from 0.75–1.5 inches (1.8–3.7 cm) in length. The trunk and branches of young wild trees are often armored with sharp, heavy thorns that can be 2–4 inches (5–10 cm) long. Mature trees typically develop a graceful, upright form with a spreading crown. The leaves turn clear yellow in fall.

FLOWERS AND FRUIT: Honey locust is a dioecious species; both male and female individuals produce relatively inconspicuous, insect-pollinated flowers in late spring. Females typically produce heavy crops of twisted, flat green pods that turn deep reddish brown at maturity. The pods, which range in size from 6–18 inches (15–45 cm) long by 1 inch (2.5 cm) wide, typically remain on the tree through the winter. They contain numerous hard, brown seeds embedded in a sticky, greenish pulp that tastes slightly sweet and is the origin of the plant's common name.

GERMINATION AND REGENERATION: The hard seeds require some form of mechanical scarification (e.g., passing through the digestive system of animals that feed on the pods) to germinate; established plants sometimes produce root suckers.

HABITAT PREFERENCES: Honey locust has a reputation as a tough, drought-resistant street tree, but in the urban environment spontaneous plants are usually found in sunny, bottomland sites or on recently disturbed soil. Unlike its cousin the black locust (*Robinia pseudoacacia*), honey locust does not fix atmospheric nitrogen and does not grow as well in nutrient-poor soils.

CULTURAL SIGNIFICANCE: Most of the honey locust trees being grown as street trees today are thornless male selections (var. *inermis*) that are less messy and less threatening than are thorny, seed-grown trees—in essence, the urban version has been neutered and declawed. In the past, farmers would leave wild trees standing so livestock could feed on the nutritious pods once they fell to the ground.

RELATED SPECIES: Silktree or mimosa (*Albizia julibrissin* Durazz.) is a medium-sized, deciduous tree up to 30 feet (9 m) tall with a broad, spreading growth habit and is native to parts of central Asia (Iran) and eastern China. It produces alternate, twice-compound leaves—up to 20 inches (50 cm) long with numerous tiny leaflets—and white or pink "powder-puff" flowers that bloom throughout the summer. Blossoms are followed by 4–6 inch (10–15 cm) long pods that are initially green but turn brown in fall. Silktree is surprisingly common in urban vacant lots and the edges of degraded woodlands as far north as New York City. It grows best in full sun and, because of its nitrogen-fixing ability, tolerates poor soil. Silktree germinates readily from scarified seed, and established trees produce root suckers, especially following winter damage or disease.

Stout spines on 1-year-old honey locust twigs

Honey locust grown from seed with stout spines on its trunk

Variation in honey locust leaves: twice compound (*left*); once compound (*center*); both once and twice compound (*right*)

Honey locust foliage and seedpods

Silktree in flower

Silktree flowers, seedpod, and foliage

Robinia pseudoacacia L.   Black Locust

Synonyms: common locust, yellow locust, white locust, false acacia, treenail

Life Form: deciduous tree; 30–70 feet (9–21 m) tall

Place of Origin: eastern North America

Vegetative Characteristics: Old black locusts have gray to brown, deeply furrowed bark that gives the trunk a rugged appearance; young stems are armed with pairs of sharp spines at the leaf nodes. Alternate, pinnately compound leaves are 8–14 inches (20–35 cm) long, with 6–20 rounded, pale blue-green leaflets that either remain the same color or turn light yellow before dropping in the fall. The plant is late to leaf out in the spring.

Flowers and Fruit: Black locust produces 6–12 inch (15–30 cm) long chains (racemes) of showy, extremely fragrant flowers in mid-to-late spring; the pea-like white flowers are approximately an inch (2.5 cm) wide with a bright yellow center and are pollinated mainly by bees. The flat brown pods are 2–4 inches (5–10 cm) long and stay on the plant well into winter.

Germination and Regeneration: Black locust seeds will germinate in sunny, dry sites, but most young sprouts are root suckers produced by established plants. Over time, a single individual can give rise to a thicket of stems that, if unchecked, will develop into a genetically uniform stand of trees.

Habitat Preferences: Black locust tolerates a wide range of soils—from saline roadsides and acidic mine tailings to rich, loamy limestone—but its greatest competitive advantage is on dry and sandy sites in full sun. In the urban environment, it is common in minimally maintained public parks, vacant lots, waste dumps, emergent woodlands, unmowed highway banks, and railroad rights-of-way. Nitrogen-fixing root nodules contain symbiotic *Rhizobium* bacteria that allow black locust to generate its own fertilizer on low-nutrient sites.

Cultural Significance: Black locust has been widely cultivated since the mid-1700s as an ornamental and as a source of rot-resistant wood for fence posts and poles. It was one of the first American trees to be widely planted on degraded soils in Europe during the late 1700s and has naturalized throughout much of that continent. It has also been planted throughout temperate Asia, where it has also naturalized. Black locust was used as a street tree in New York City as early as 1751. In the species' native Appalachia, black locust timbers were used as supports in coal mines because of their great tensile strength and rot resistance. The leaves, seeds, and bark of the plant are toxic to both humans and livestock, but the flowers are edible either raw or cooked. Numerous horticultural varieties and hybrids of *Robinia* have been developed with unusual growth habits and colorful foliage. Black locust is widely planted on disturbed reclamation sites, including landfills and mine tailings. The flowers are extremely attractive to bees and produce an excellent honey.

Stand of mature
black locust in
Riverside Park,
New York City

Black locust in bloom in late spring

Black locust
flowers

Black locust foliage

Stand of black locust stems developed from
root sprouts

Stout thorns on black
locust stems

Quercus palustris Müchh.　Pin Oak

Synonyms: none

Life Form: deciduous tree; 60–70 feet (18–21 m) tall

Place of Origin: eastern North America

Vegetative Characteristics: Young pin oaks develop a pyramidal growth habit with a prominent central leader that is carried up to the top of the crown and distinctly horizontal side branches. On older trees the dead lower branches often persist and form a pendulous "skirt" around the trunk. Shiny, green, alternate leaves are 3–6 inches (7.5–15 cm) long by 2–4 inches (5–10 cm) wide and have 5–7 sharp-pointed lobes that are separated by deep, U-shaped sinuses or indentations; their fall color varies from brown to bronze to deep red. The bark is grayish brown and smooth; with age it develops shallow ridges and furrows. Young trees typically hold onto their withered, brown leaves through the winter, an important aid to identification. Pin oak has a shallow, fibrous root system.

Flowers and Fruit: Delicate racemes of light yellow, wind-pollinated flowers are produced in the axils of the leaves in mid-spring. The relatively small acorns are up to 0.75 inch (1.8 cm) across with light longitudinal markings; they mature at the end of their second growing season.

Germination and Regeneration: Squirrels and birds consume and disperse the acorns when they ripen in October. Acorns that escape predation germinate the following spring. Saplings resprout vigorously from the base following injury to the stem or its displacement from the vertical orientation.

Habitat Preferences: As a spontaneous tree in the urban environment (usually escaped from cultivation), pin oak is common along the edges of minimally maintained open space, especially the margins of freshwater wetlands, ponds, and streams. It grows best with acidic soil conditions and can tolerate sites that are flooded in winter (when it is dormant) and dry in summer.

Ecological Functions: Heat reduction in paved areas; tolerant of roadway salt and compacted soil; food and habitat for wildlife; erosion control on slopes; stream and riverbank stabilization.

Cultural Significance: Pin oak is a widely planted street tree in the East and, despite its preference for bottomland habitats, is quite drought-tolerant. It also survives transplanting better than most other oaks. As a cultivated specimen in a park or on the street, this species has few peers in terms of adaptability, durability, and attractiveness.

Pin oak growth habit

Pin oak foliage

Fall foliage of a mature pin oak in a Boston wetland

Young pin oaks hold onto their withered leaves throughout the winter

Pin oaks in late autumn

Pin oak acorns

Quercus rubra L.   Red Oak

SYNONYM: northern red oak

LIFE FORM: **deciduous tree**; 60–90 feet (18–28 m) tall

PLACE OF ORIGIN: eastern North America

VEGETATIVE CHARACTERISTICS: The bark of a mature red oak consists of rough brown or black ridges separated by smooth, light patches of trunk. Alternate, dark green leaves have a glossy appearance, are approximately 6 inches (15 cm) long by 4 inches (10 cm) wide, and have 7–11 bristle-tipped lobes separated by U-shaped indentations that extend roughly halfway to the main vein. Red oaks typically develop a single thick trunk and a rounded crown with wide-spreading branches. The foliage is late to color up in the fall and turns anywhere from deep red to yellow-brown depending on the weather.

FLOWERS AND FRUIT: Red oak produces delicate, dangling chains—up to 5 inches (12.5 cm) long—of light yellow, wind-pollinated flowers in mid-spring, when the leaves are half grown and still yellow-pink. In the fall the plant produces 1 inch (2.5 cm) long, medium brown acorns with a shallow, saucer-like cap that covers only the base of the nut.

GERMINATION AND REGENERATION: Heavy crops of acorns (known as *mast*) are produced every 3–5 years; they are eagerly consumed by birds and mammals and are greedily collected and buried by squirrels as part of their winter food supply; uneaten seeds germinate in spring. Saplings growing in the forest understory resprout readily following injury, producing 3–4 inch (7.5–10 cm) tall "seedlings" that can be 5 or 10 years old. Large trees can also resprout following damage to their main trunk.

HABITAT PREFERENCES: Red oak is a relatively fast-growing, upland species that is widely cultivated along streets and in parks. It grows best in acid soils and full sun but is tolerant of air pollution, drought, and shade. It reproduces spontaneously in disturbed urban woodlands and along roadway edges.

ECOLOGICAL FUNCTIONS: Heat reduction in paved areas; food and habitat for wildlife; erosion control on slopes; stream and riverbank stabilization.

CULTURAL SIGNIFICANCE: Red oak is widely planted as a shade tree in urban environments; its lumber is very valuable, and its tannin-rich foliage deters feeding by insects, not including gypsy moths. Native Americans used the acorns to make bread, but only after boiling them to remove the bitter taste. The tree is widely planted throughout Europe.

Spontaneous red oak street tree in Boston

Red oak flowers and expanding spring foliage

Red oak bark

Red oak acorns

Red oak foliage

Juglans nigra L.   Black Walnut

SYNONYMS: none

LIFE FORM: **deciduous tree**; up to 60–90 feet (18–28 m) tall

PLACE OF ORIGIN: eastern North America

VEGETATIVE CHARACTERISTICS: Black walnut is a coarsely branched tree that forms a tall, straight trunk when allowed to grow freely. On young trees, the light brown bark is scaly; on older trees, bark is dark brown to black and deeply furrowed. Alternate, pinnately compound leaves are 12–24 inches (30–60 cm) long with 12–24 short-stalked, slightly serrated leaflets that give off a pungent, spicy scent when crushed. The foliage turns an unremarkable yellow to brown in the fall. Thick winter twigs have a large terminal bud and prominent U-shaped leaf scars.

FLOWERS AND FRUIT: Black walnut produces wind-pollinated flowers in May along with its new leaves; pendulous male catkins are 3–5 inches (7.5–12.5 cm) long at maturity; separate female flowers are produced in clusters of 1–3 higher up on the same twig. Distinctive, spherical fruits, 2 inches (5 cm) in diameter, mature in October. They are green when shed from the tree, with a fleshy husk surrounding a deeply grooved, thick-shelled nut. Fruits turn yellow and then black as they rot; picking them up at this stage will leave one's hands deeply stained.

GERMINATION AND REGENERATION: Black walnuts are a favorite food of squirrels, who collect and bury the nuts for their winter food. Germinating seedlings produce a vigorous taproot that makes them hard to transplant. Saplings, and to a lesser extent mature trees, will sprout vigorously from the base following injury. Groves of young trees often develop around mature, cultivated specimens.

HABITAT PREFERENCES: Black walnut grows rapidly in deep, moist soil in full sun. It does not grow well in soils that are overly dry, wet, or acidic. Black walnut grows surprisingly well under urban conditions and, if a mature tree is in the neighborhood, seedlings will spring up in adjacent properties (thanks to the squirrels).

CULTURAL SIGNIFICANCE: Heavy, dark black walnut wood is extremely valuable if it comes from trees with a tall, straight trunk that can be peeled to make veneer. The nuts are edible but require a lot of work to crack open; their fleshy husks have been used to dye cloth brown. The tree was once widely planted as an ornamental but is seldom used today, mainly because of its "messiness" and because its roots produce a chemical (juglone) that suppresses the growth of other nearby plants, including tomatoes.

RELATED SPECIES: **Bitternut hickory (*Carya cordiformis* (Wangenh.) K. Koch)** is one of the few hickories that reproduce spontaneously in the urban environment. Its compound leaves are 6–10 inches (15–25 cm) long with 7–11 serrated, narrow leaflets. In October, the leaves turn bright yellow and the tree drops an abundance of 1 inch (2.5 cm) long fruits consisting of a fleshy, segmented husk surrounding a thin-shelled nut with prominent ridges. The nuts have a bitter taste, but that does not discourage squirrels from eating them. In winter, the tree can be readily identified by its naked, sulfur-yellow buds.

A stand of black walnuts planted by Boston squirrels

The naked, sulfur-yellow buds of bitternut hickory make it easy to identify in winter

Black walnut foliage

Black walnuts with and without their husks

A mature bitternut hickory in fall color in Boston

Bitternut hickory nuts with and without their husks

Morus alba L.   White Mulberry

SYNONYM: common mulberry

LIFE FORM: **deciduous tree**; 30–60 feet (9–18 m) tall

PLACE OF ORIGIN: eastern Asia

VEGETATIVE CHARACTERISTICS: The bark of young mulberry stems is light gray in color, often with a slight orange or yellowish tint; bark on older trees is brown and furrowed. The simple leaves are alternate, glossy, and 2–6 inches (5–15 cm) long. They have coarsely toothed margins and are highly variable in shape, ranging from unlobed to 5 or 6 lobes separated by deep sinuses; indeed, this heterogeneity in leaf shape is a key identification feature. The young twigs exude milky sap when broken, the leaves turn a clear yellow in fall, and the outer bark of the roots is a dull orange color.

FLOWERS AND FRUIT: White mulberry produces inconspicuous greenish-yellow, wind-pollinated flower spikes (*catkins*) in early spring. Male and female flowers are produced separately on the same tree or on different trees. Fruits are very juicy and turn black or purple (occasionally pink or white) at maturity. Roughly the size and shape of a blackberry, they ripen in mid-to-late summer; they are edible but rather insipid.

GERMINATION AND REGENERATION: White mulberry fruits are readily consumed and dispersed by birds, and seedlings readily sprout up beneath their roosting sites. Damaged trees resprout vigorously from the base of the trunk.

HABITAT PREFERENCES: White mulberry is a highly adaptable, drought-tolerant species that grows best in full sun. In the urban environment, it is common in minimally maintained public parks, disturbed or emergent woodlands, chain-link fence lines, stream and riverbanks, unmowed highway banks and median strips, and sidewalk and foundation cracks.

ECOLOGICAL FUNCTIONS: Heat reduction in paved areas; tolerant of roadway salt and compacted soil; food and habitat for wildlife; erosion control on slopes.

CULTURAL SIGNIFICANCE: White mulberry has been cultivated for several thousand years in China for its leaves, which are fed to silkworm caterpillars. In addition, the fruit is eaten raw, a medicinal tea is made from the leaves after they fall from the tree, and the bark of the root is used for a variety of antimicrobial purposes. White mulberry was introduced into North America in the 1820s in the hope of establishing a silkworm industry, but that never materialized. The dwarf variety *multicaulis* was particularly popular to the extent that Darlington and Thurber (1859, 295) reported "a sort of *Multicaulis monomania* (or Moromania!)—so universal, and engrossing, that it became absolutely ludicrous; and was scarcely exceeded in absurdity by the nearly contemporaneous epidemic which afflicted the nation in reference to its financial concerns. Almost everybody was eagerly engaged in cultivating myriads of trees . . . without stopping to enquire where they could be sold, or who would be likely to buy!" With the help of birds, white mulberry has become naturalized in a wide variety of urban and rural habitats throughout the Northeast.

A mature white mulberry on a Detroit street

White mulberry growth form

White mulberry seedlings under stress often produce small leaves

White mulberry leaf shape is highly variable

White mulberry fruits at various stages of maturity

Fraxinus pennsylvanica Marshall   Green Ash

SYNONYM: red ash

LIFE FORM: **deciduous tree**; 30–60 feet (9–18 m) tall

PLACE OF ORIGIN: eastern North America

VEGETATIVE CHARACTERISTICS: Green ash produces opposite, pinnately compound leaves, 10–12 inches (25–30 cm) long, composed of 7–9 lance-shaped leaflets plus a single terminal leaflet. Smooth, yellowish-green leaves come out late in the spring and turn a clear, bright yellow in fall. Branches are opposite one another, and the brown to gray bark is covered with shallow, vertical fissures.

FLOWERS AND FRUIT: Green ash produces compact 1–2 inch (2.5–5 cm) long clusters of wind-pollinated, unisexual flowers in spring, either just before or with the leaves; individual trees are either female (with seeds) or male (seedless).

GERMINATION AND REGENERATION: Wind-dispersed, winged fruits (samaras) mature in late September and often remain on the plant through the winter. They germinate in spring under a wide variety of conditions; young trees sprout vigorously from the base following injury.

HABITAT PREFERENCES: Green ash is a light-demanding species that grows best in nutrient-rich soils that characterize seasonally flooded river floodplains and stream banks. It is highly adaptable and can tolerate inundation in spring and drought in summer. Green ash grows in disturbed urban sites where water accumulates, including roadside drainage ditches, fence lines, and highway medians.

ECOLOGICAL FUNCTIONS: Heat reduction in paved areas; food and habitat for wildlife; erosion control on slopes; and bank stabilization along streams.

CULTURAL SIGNIFICANCE: Because of its tolerance of both wet and dry conditions, green ash—especially seedless (male) cultivars—has been widely planted as a street tree in urban areas throughout the Northeast and Midwest, where it has also been extensively used for shelterbelts. The species has been widely planted in Europe and Asia and has escaped cultivation in many countries. In the 1990s, the emerald ash borer (*Agrilus planipennis*) from eastern Asia was inadvertently introduced into North America (Michigan) and has since spread south and east killing both wild and cultivated ash trees it encounters. The situation has gotten so bad that people have mostly stopped planting any *Fraxinus* species.

RELATED SPECIES: **White ash (*Fraxinus americana* L.)** is a much larger, native upland tree than the bottomland green ash, reaching up to 90 feet (28 m) tall under optimal conditions. It has a moderate growth rate and does best in moist, sunny situations, but its wind-dispersed seeds can colonize drier sites if the soil is reasonably good. Its opposite, compound leaves have 7–9 leaflets, are 8–12 inches (20–30 cm) long, and turn an unusual purple color in fall. The wood is very straight-grained and has been used for making tool handles, hockey sticks, and baseball bats.

Green ash in fall color in Boston

Bark of a mature white ash

White ash in fruit

Green ash
winter
twig

New foliage of white ash is often purple

Green ash seedling

Ligustrum obtusifolium Sieb. & Zucc.   Border Privet

SYNONYMS: *Ligustrum amurense*, Amur privet

LIFE FORM: **semi-evergreen shrub**; 9–12 feet (3–4 m) tall

PLACES OF ORIGIN: eastern Asia

VEGETATIVE CHARACTERISTICS: Border privet is a multistemmed shrub with arching branches; its opposite, glossy, dark green leaves are lanceolate to oblong in shape, 1–2 inches (2.5–5 cm) long by 0.5–1 inch (1.2–2.5 cm) wide. The upper leaf surface is glaucous, whereas the midrib on the lower surface is pubescent. In late fall, border privet leaves turn purplish-black before falling off in November or December.

FLOWERS AND FRUIT: Border privet produces tubular, white, insect-pollinated flowers in June that emit a distinctive scent that some people find unpleasant. They are 0.4 inch (1 cm) long with recurved petals on panicles up to 2 inches (5 cm) long. Small, round, black fruits mature in October and often persist into winter.

GERMINATION AND REGENERATION: Birds readily consume border privet fruits in fall and deposit the seeds beneath their roosting sites; seedlings establish best in moist, shady or sunny locations. Established plants sprout readily from the base with or without injury.

HABITAT PREFERENCES: Border privet grows in a variety of conditions from sunny stream banks to shady forest understories. It is presently more common in suburban and rural habitats than in cities, but that could change in the future. As a cultivated hedge plant, border privet is tolerant of both drought and road salt.

ECOLOGICAL FUNCTIONS: Food and habitat for wildlife; erosion control along rivers and streams.

CULTURAL SIGNIFICANCE: Border privet was probably introduced into North America from Japan in 1875 by Thomas Hogg as an ornamental species and was used primarily for hedging purposes (Del Tredici 2017a).

RELATED SPECIES: **European privet (*Ligustrum vulgare* L.)** was introduced into North America during the late 1700s for medicinal purposes and for creating hedges. Darlington and Thurber (1859, 266) noted that the plant "has become completely naturalized, and is found plentifully in New England, New York and Pennsylvania." European privet is still cultivated as a hedge plant in urban areas because of its tolerance of severe pruning, neglect, and road salt, but its use seems to be diminishing as homeowners, tired of shearing it once or twice a year, replace it with plastic fencing. European privet does not seem to escape into natural areas as readily as does border privet.

Border privet flowers

Border privet in fruit

Border privet
fall color

Border privet growth habit in
November

European privet is a common hedge
plant in cities

European privet foliage and immature
fruits

Paulownia tomentosa Steud.   Princess Tree

SYNONYMS: *Paulownia imperialis*, empress tree, karri-tree, royal paulownia

LIFE FORM: deciduous tree; up to 60 feet (18 m) tall

PLACE OF ORIGIN: temperate East Asia

VEGETATIVE FEATURES: *Paulownia* is a fast-growing, sparsely branched tree with stout, pithy twigs. On mature trees, the opposite, heart-shaped leaves are 6–12 inches (15–30 cm) long and nearly as wide, and are covered with velvety hairs, especially on the undersides; on juvenile trees, the leaves can be up to 2 feet (60 cm) across with 2 small secondary lobes. Leaves often fall while still green or turn brown after experiencing a hard freeze in autumn.

FLOWERS AND FRUIT: Prominent clusters of fuzzy brown flower buds develop at the ends of the branches in the fall. Buds remain in a rudimentary state of development throughout the winter, then expand in April or May to produce spectacular 2 inch (5 cm) long tubular flowers. The flowers, which are pale violet with dramatic yellow stripes on the inside of the corolla, are pollinated mainly by bees. They are followed by pointed, pecan-shaped woody capsules approximately 1.25–2 inches (3–5 cm) long that split open to release hundreds of small, wind-dispersed seeds in the fall. The spent pods often remain on the tree for several years, a useful feature that helps with identification. A mature *Paulownia* tree can produce up to 20 thousand seeds per year.

GERMINATION AND REGENERATION: Seeds germinate in spring on bare ground. Mature plants typically produce root suckers, especially following damage to the primary trunk. These shoots can arise at a considerable distance from the trunk, and most people assume they are seedlings.

HABITAT PREFERENCES: Princess tree is a light-demanding, drought-tolerant plant that grows in a variety of disturbed urban habitats, including vacant lots, chain-link fence lines, pavement and masonry cracks, rock outcrops, and highway and railroad banks. This species is currently most abundant in the mid-Atlantic region but can be expected to move farther north as the climate becomes warmer.

ECOLOGICAL FUNCTIONS: Heat reduction in urban areas; tolerant of roadway salt and compacted soil; erosion control on slopes; food and habitat for wildlife.

CULTURAL SIGNIFICANCE: *Paulownia* was introduced into North America from Europe in 1844. Its rapid spread throughout the East Coast was supposedly facilitated when seeds used as packing material to protect imported Chinese porcelain were discarded. The species was hyped in Sunday newspaper supplements in the 1970s and '80s as a "wonder tree" that grows 6 feet (1.9 m) a year. Its light, fine-grained wood is highly valued in Japan for making a number of special items.

Princess tree has taken over the front yard of an abandoned house in West Philadelphia

Growth habit of a spontaneous princess tree in New London, Connecticut

Princess tree saplings growing in good conditions produce huge leaves

Princess tree flowers

Winter twig of princess tree showing spent fruits and immature flower buds ready to open in spring

Clematis terniflora DC.   Sweet Autumn Clematis

Synonyms: *Clematis paniculata, C. disocoreaefolia, C. maximowicziana,* yamleaf clematis

Life Form: deciduous vine (liana); stems can grow up to 20 feet (6 m) in length

Place of Origin: northeast Asia

Vegetative Characteristics: Unlike high-climbing vines that ascend their host trees by means of twining stems, rootlets, or tendrils, sweet autumn clematis uses touch-sensitive leaf petioles. This means that the plant can attach only to small-diameter objects and has more of a sprawling than a climbing growth habit; nevertheless, it can easily cover trellises, fences, and shrubs. Its opposite, pinnately compound leaves are composed of 3–5 dark green leaflets that remain on the plant until a hard frost arrives. In areas with mild winters, sweet autumn clematis will hold onto its leaves until spring.

Flowers and Fruit: Sweet autumn clematis produces large masses of delight-fully scented flowers for two weeks or more between late August and October, depending on latitude. Individual white, star-like flowers have 4 petals and are approximately 1.25 inches (3 cm) across; at peak bloom, the entire plant forms a spectacular blanket of white sitting atop whatever it is growing on. Seeds are pro-duced in clusters of five, each with a "tail" covered with long, silky hairs.

Germination and Regeneration: Seeds are dispersed by wind in late autumn and germinate the following spring, typically under conditions of partial shade.

Habitat Preferences: Like most other *Clematis* species, sweet autumn grows best with full sun on its leaves and its root system in the shade. Unlike many other clematis, however, it is not finicky about soils and grows well in disturbed wood-land edges, especially along streams and rivers.

Cultural Significance: Sweet autumn clematis was introduced as an ornamen-tal into North America from Japan by Thomas Hogg in 1875 (Del Tredici 2017a) and quickly became popular because of its abundant September flowers, vigorous growth, and ease of cultivation. It was first reported growing spontaneously in the 1950s and has been increasing steadily ever since. It is now considered invasive in the mid-Atlantic region and parts of the Northeast.

Related Species: Virgin's bower (*Clematis virginiana* L.) is a woody vine native to eastern North America. Its compound, matte-green leaves are coarsely toothed and clearly distinct from the glossy, entire leaves of sweet autumn clematis. It pro-duces large masses of 4-petaled, white flowers from July through September fol-lowed by fluffy, wind-dispersed seeds in October. Each seed head contains 10–20 achenes as opposed to 4–12 for sweet autumn clematis. Virgin's bower is more common in rural areas than in the city, especially in moist roadside ditches and hedgerows.

Cultivated sweet autumn clematis in full bloom in Boston

Flowers of sweet autumn clematis

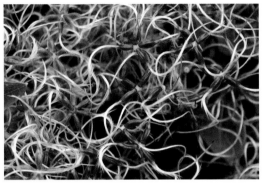

Fruits of sweet autumn clematis

Virgin's bower foliage and fruits

Virgin's bower flowers

Close-up of virgin's bower fruits

Frangula alnus P. Mill.   Glossy Buckthorn

SYNONYMS: *Rhamnus frangula*, alder buckthorn, black dogwood

LIFE FORM: **deciduous shrub** or **small tree**; 6–16 feet (2–5 m) tall

PLACE OF ORIGIN: Europe, North Africa, and central Asia

VEGETATIVE CHARACTERISTICS: Glossy buckthorn has smooth gray bark that is covered with prominent white or gray lenticels; it typically develops a multi-stemmed growth habit. Alternate, glossy green leaves are 1–2.5 inches (2.5–6.2 cm) long, round to oval in shape, and have prominent parallel veins and entire margins. Leaves maintain their dark green color well into fall, long after native species have gone dormant. Fall color, when it finally appears, is dull yellow. Roots are a distinctive maroon color.

FLOWERS AND FRUIT: Glossy buckthorn produces inconspicuous, solitary, 5-petaled, yellow-green flowers in the leaf axils throughout the summer; they are pollinated by bees. Flowers develop into 0.25 inch (6 mm) round fruits that are initially green and then turn red before ripening to black from July through September. The simultaneous presence of flowers and multicolored fruits is a key feature that helps to distinguish glossy buckthorn from common buckthorn (*Rhamnus cathartica*). Among woody plants, glossy buckthorn is highly unusual in ripening its fruits asynchronously over a period of several months.

GERMINATION AND REGENERATION: The fruits are readily consumed and dispersed by birds, and the seeds germinate under a wide variety of conditions from sunny and dry to shady and moist. Secondary stems sprout spontaneously from the base.

HABITAT PREFERENCES: Glossy buckthorn grows best in humus-rich soils and is abundant in the understory of minimally maintained woodlands; along the margins of freshwater wetlands, streams, and rivers; in vacant lots and rubble dumps; along chain-link fences; and on roadway edges and railroad rights-of-way.

ECOLOGICAL FUNCTIONS: Nutrient absorption in wetlands; tolerant of roadway salt; food and habitat for wildlife; erosion control on slopes.

CULTURAL SIGNIFICANCE: The stem bark of glossy buckthorn has a long history of medicinal use in Europe as a laxative and to induce vomiting. The species was introduced into North America prior to 1800, probably for medicinal purposes. It has been spreading on its own since the early 1900s and has dramatically increased its range since the 1950s, particularly along the edges of wetlands. Many northeastern and Midwestern states list glossy buckthorn as an invasive species.

Glossy buckthorn foliage and fruits

Glossy buckthorn leafing out early in a woodland understory

Glossy buckthorn flowers

Multistemmed growth form of glossy buckthorn

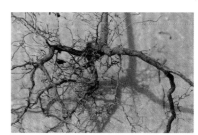

Red roots of glossy buckthorn

Glossy buckthorn fruits mature continuously throughout the growing season

Rhamnus cathartica L.   Common Buckthorn

SYNONYMS: red root, hartshorn, waythorn, highwaythorn, ramsthorn, purging buckthorn

LIFE FORM: **deciduous shrub** or **small tree**; 10–25 feet (3–8 m) tall

PLACE OF ORIGIN: Eurasia and North Africa

VEGETATIVE CHARACTERISTICS: In cultivation, common buckthorn may have a single trunk, but as a spontaneous plant it is usually multistemmed. The outer bark is gray, the inner bark is bright yellow, and the twigs are tipped with sharp thorns. Simple, glossy, dark green leaves are mostly opposite with finely serrated edges; they are approximately 1.5–3 inches (3.7–7.5 cm) long and are elliptical to egg-shaped. Plants leaf out early in spring and in the fall retain their green leaves much longer than the surrounding native vegetation—falling only after they experience a hard frost.

FLOWERS AND FRUIT: Common buckthorn produces inconspicuous clusters of greenish-yellow, insect-pollinated flowers in spring. Clusters of small black fruits, about 0.25 inch (6 mm) wide, mature simultaneously in September.

GERMINATION AND REGENERATION: Fruits are readily consumed and dispersed by birds, and seeds germinate under a wide variety of conditions; mature plants sprout readily from the base.

HABITAT PREFERENCES: Common buckthorn grows best in partially shaded woodland edges in soil with a neutral pH. It is also common along the margins of freshwater wetlands, ponds, and streams; in degraded woodlands; and along roadsides.

ECOLOGICAL FUNCTIONS: Tolerant of roadway salt and compacted soil; food and habitat for wildlife; erosion control on slopes.

CULTURAL SIGNIFICANCE: Common buckthorn was included in Dioscorides' first-century herbal, *De Materia Medica*. Its bark and, more commonly, its berries have long been used in traditional European medicine to accelerate defecation (hence the name *cathartica*), to induce vomiting, and to treat gout, dropsy, and rheumatism. Unripe berries were also used to make a dye known as sap-green. Common buckthorn was introduced into North America during the 1700s, for both medicinal and horticultural purposes. The naturalist Thomas Nuttall vividly described one of these uses in 1846 (p. 52): "In New England, particularly in the vicinity of Boston, this species is much employed for useful and ornamental hedges, and bearing well to be cut, growing thick, and remaining green til winter, it is strongly recommended for this useful purpose." Common buckthorn escaped from cultivation as early as 1811 in the mid-Atlantic region and many states—especially in the Midwest—now list it as an invasive species.

Growth habit of a mature common buckthorn shrub

Common buckthorn along a stream in Boston is still green in late fall

Common buckthorn flowers in spring

Common buckthorn foliage stays green after the Norway maple and pin oak surrounding it have turned

Common buckthorn fruits and foliage

SPECIES: The identification of spontaneous apple trees is usually uncertain given the hybrid nature of most eating and ornamental apple varieties. Field manuals typically describe naturalized plants as belonging to the common edible apple group *Malus pumila* P. Miller, but **Siberian crabapple** (*M. baccata* (L.) **Borkh.**) and **Toringo crabapple** (*M. sieboldii* (Regal) **Rehder**) are also common naturalized species in the Northeast.

LIFE FORM: deciduous tree; 15–40 feet (5–12 m) tall

PLACE OF ORIGIN: Eurasia

VEGETATIVE CHARACTERISTICS: Wild apples typically grow as small, multi-stemmed trees or shrubs, usually broader than tall, with rough, scaly bark. Alternate, simple leaves are 2–4 inches (5–10 cm) long by 1–2 inches (2.5–5 cm) wide; they can be entire, serrated, lobed, or deeply incised with varying degrees of pubescence, and they typically turn yellow in the fall.

FLOWERS AND FRUIT: Wild apples produce showy white or pink, insect-pollinated flowers in early spring; they are produced in clusters on short shoots and have 5 petals that are positioned above the ovary. Small fruits are produced in the fall and range in diameter from 0.25–1 inch or so (0.6–2.5 cm); they can be red, yellow, or green at maturity.

GERMINATION AND REGENERATION: Fruits are eagerly consumed by birds and mammals, and they may deposit the seeds beneath their roosting sites some distance from the tree. Seeds germinate in a variety of conditions from sun to shade; saplings readily regenerate new shoots following damage. Some species produce root suckers and form thickets.

HABITAT PREFERENCES: As a cultivated plant, apples grow best in loamy, well-drained, acidic soils in full sun; it can survive in understory shade but will produce few flowers and little fruit. As a spontaneous plant, apples are common along the edges of disturbed woodlands and minimally maintained public parks, on rock outcrops and stone walls, and along unmowed highway banks and railroad tracks.

ECOLOGICAL FUNCTIONS: Heat reduction in paved areas; tolerant of roadway salt and compacted soil; food and habitat for wildlife; erosion control on slopes.

CULTURAL SIGNIFICANCE: Numerous species, hybrids, and cultivars of species in the genus *Malus* are widely cultivated for consumption (both eating and drinking) and for ornamental purposes. A few species of apple are native to central North America, but most edible apple varieties were introduced from Europe, some arriving with the first settlers. Flowering crabapples arrived in the 19th and 20th centuries, initially from Europe and later from Asia. Spontaneous crabapples are mostly a mongrel race of trees and some individuals occasionally produces fruits large enough to make jelly.

Wild apple growing on the old High Line in New York City

A wild apple in full flower in Detroit

Growth habit of a wild apple seedling

Wild apple growth habit

Wild apple flowers

Wild apples

Prunus serotina Ehrh.   Black Cherry

SYNONYMS: *Cerasus serotina*, rum cherry, wild black cherry, bird cherry

LIFE FORM: **deciduous tree**; 40–75 feet (12–23 m) tall

PLACE OF ORIGIN: eastern North America

VEGETATIVE CHARACTERISTICS: Black cherry has dark bark that is distinctively marked with horizontal white lines (*lenticels*); with age the bark breaks up into rectangular plates that make it quite rough. Alternate, lance-shaped leaves are dark green and glossy with finely serrated margins; they are typically 2–5 inches (5–12.5 cm) long and approximately an inch or so (2.5 cm) wide; they turn yellow or pale pink in fall. The twigs and foliage, when crushed, smell like bitter almonds (prussic acid). Black cherry is a tall, straight-growing tree in the rich Appalachian forests, but in disturbed urban environments it is typically much smaller with a contorted, irregular form. Twigs and small branches on young trees are often infected with ugly, swollen, woody tumors caused by the black knot fungus.

FLOWERS AND FRUIT: Black cherry produces 4–5 inch (10–12.5 cm) long chains of insect-pollinated, white flowers that terminate the leafy shoots in May and early June. These are followed in August and September by dangling clusters of fleshy, purple-black cherries about 0.33 inch (8 mm) in diameter.

GERMINATION AND REGENERATION: Birds eagerly consume black cherry fruits and disperse the seeds, which germinate beneath their roosts in habitats that vary from sun to shade and moist to dry. Saplings and established plants resprout vigorously from the base following injury.

HABITAT PREFERENCES: Black cherry grows best in full sun and fertile soil but it also tolerates drought and moderate shade; it grows abundantly in dry, sandy soils near the coast. In the urban environment, black cherry is found along the margins of minimally maintained public parks and unmowed highway edges, in the understory of disturbed woodlands, and along sunny riverbanks and chain-link fence lines.

CULTURAL SIGNIFICANCE: The fruits are used to make jelly and wine ("cherry bounce"), and the rich, red heartwood is highly prized for furniture and cabinetmaking. The powdered inner bark was used in 19th-century patent medicines to treat coughs and bronchitis (Dr. Wistars's Balsam of Wild Cherry and Ayer's Cherry Pectoral were popular), and Smith Brothers' Wild Cherry Cough Drops are still sold today. Livestock that eat wilted black cherry foliage can become sick or even die from the cyanide contained in the leaves. The species was widely planted in the Netherlands and portions of central Europe from the early 1900s through the 1950s and is now considered an invasive species in those countries.

RELATED SPECIES: **Choke cherry (*Prunus virginiana* L.)** is a shrubby, root-suckering species that typically grows to approximately 15 feet (4.5 m) tall. Its dull green leaves are shorter and broader than those of black cherry and it flowers a bit earlier. Choke cherry fruits are red in summer and mature to purple in September, and they are an important food for wildlife. Its leaves turn reddish-purple in fall.

Black cherry
growth habit

Black cherry
trunk

Black cherry flowers

Black cherry foliage

Mature black cherry fruits

Choke cherry fruits

Pyrus calleryana Decne.   Callery Pear

Synonym: Bradford pear

Life Form: deciduous tree; 30–50 feet (9–15 m) tall

Place of Origin: China, Japan, and Korea

Vegetative Characteristics: Callery pear is a fast-growing, single-stemmed tree with upright branches and a pyramidal form when young, becoming broad and spreading over time. It produces glossy, dark green leaves 1.5–3 inches (3.7–7.5 cm) long by 1.5–2 inches (3.7–5 cm) wide that taper to a sharp point. Leaves are alternate, smooth, egg-shaped (*ovate*) to broadly ovate, and turn various shades of scarlet-orange to wine-red to deep purple in November, long after most other trees have shed their leaves. The smooth, grayish-brown bark eventually develops shallow fissures.

Flowers and Fruit: Callery pear produces pure white flowers with 5 broad petals in very early spring, typically before the leaves come out in March or April. A tree in full flower on a drab city street can be a spectacular sight. Individual flowers are 0.5–0.75 inch (1.2–1.8 cm) across and occur in clusters (*corymbs*) that are approximately 3 inches (7.5 cm) across. They emit an unpleasant—some would say foul—odor that attracts a variety of insects, including flies and honey bees. Small, round fruits, 0.5 inch (1.2 cm) in diameter, are covered with russet-colored dots and mature in late autumn.

Germination and Regeneration: Callery pear fruits remain on the tree until consumed by birds—notably robins and European starlings—that disperse the seeds far and wide. When its main trunk has been damaged, the root system can give rise to a dense thicket of new stems, often covered with sharp spur shoots.

Habitat Preferences: Frank Meyer, who introduced the species from China in 1918, provided an amazing description of the plant's native habitat: "*Pyrus calleryana* is simply a marvel. One finds it growing under all sorts of conditions; one time on dry, sterile mountain slope; then again with its roots in standing water at the edge of a pond; sometime in open pine forest, then again among scrub on blue-stone ledges in the burning sun; sometimes in low bamboo-jungle and then again along the course of a fast flowing mountain stream or on the occasionally burned-over slope of a pebbly hill" (Culley and Hardiman 2007, 960).

Cultural Significance: Callery pear was introduced into North America from China in 1918 as part of an effort by the USDA to breed fire blight resistance into the European pear. In 1952, a non-spiny plant from the original collections was recognized as distinct—the cultivar 'Bradford'—and propagated. It became commercially available in 1962 and was immediately recognized as the perfect landscape tree—beautiful flowers, upright growth habit, disease resistant, and drought tolerant. Over the course of the next 20 years, new selections of Callery pear were introduced and the incidence of cross-pollination and fruit production increased dramatically. It was not until the 1990s that people recognized the species' invasive potential in the mid-Atlantic region (Culley and Hardiman 2007). Although the spontaneous spread of Callery pear in the Northeast is not as bad as it is farther south, it is definitely starting to happen in our region, facilitated by climate change and hungry birds.

Spontaneous Callery pear in flower in March along the
Anacostia River in Washington, DC

Callery pear leaves

Callery pear flowers

Callery pear fruits, approximately
one-half inch in diameter

Spontaneous Callery pear in fall
color during November in an
abandoned agricultural field

Callery pear in bloom in April in New York
City

Rosa multiflora Thunb.   Multiflora Rose

SYNONYMS: Japanese rose, living fence, rambler rose

LIFE FORM: **climbing** or **scandent shrub**; stems 10–16 feet (3–5 m) long

PLACE OF ORIGIN: temperate East Asia

VEGETATIVE CHARACTERISTICS: Stems of multiflora rose are smooth, green or reddish, and bear vicious, recurved thorns that readily catch hold of clothing or skin. Alternate, pinnately compound leaves have 5–11 leaflets that are 0.5–1.5 inches (1.2–3.7 cm) long with serrated edges and fringed stipules at their base; they retain their lustrous bright green coloration until late in the fall, finally turning pale yellow. While it is not really a vine, the arching stems of multiflora rose show a strong tendency to grow up through other plants and smother them. Under ideal growing conditions, it forms impenetrable thickets.

FLOWERS AND FRUIT: In May and June, multiflora rose is covered with pyramidal clusters of fragrant, white to pale pink, insect-pollinated flowers that are 1–1.5 inches (2.5–3.7 cm) wide with 5 notched petals and a yellow center. Flowers are followed by loose bunches of small, round or egg-shaped, dull red rose hips that are 0.25 inch (6 mm) long; the rose hips mature in late summer and remain on the plant through most of the winter.

GERMINATION AND REGENERATION: Fruits are consumed and dispersed by birds, and the seeds can germinate under a wide variety of soil and light conditions. They also can remain viable in the soil seed bank for up to 20 years. Established plants sprout prolifically from the perennial woody base and the tips of the arching stems root where they touch the ground, allowing plants to spread across the landscape.

HABITAT PREFERENCES: Multiflora rose grows in a wide variety of ecological conditions and soil types ranging from moist, sunny sites to dry, shady ones. It is common in vacant lots, rubble dumps, woodland edges, abandoned meadows, stream and riverbanks, and along chain-link fences and railroad tracks.

ECOLOGICAL FUNCTIONS: Tolerant of roadway salt and compacted soil; food and habitat for wildlife; erosion control on slopes.

CULTURAL SIGNIFICANCE: Multiflora rose was introduced from China into Europe in the 1860s and sent from Europe to North America—to the Arnold Arboretum in Boston—in 1874 (Sargent 1890). Between the 1930s and 1960s, the plant was widely promoted and distributed by the U.S. Soil Conservation Service for erosion control under the name "living fence," and small-scale farmers were encouraged to plant it for hedgerows (Elton 1958). With the help of birds, multiflora rose now dominates abandoned fields and the margins of disturbed woodlands throughout the Northeast. It is listed as an invasive species in more than 30 states.

Multiflora rose growth and flowering habit

Multiflora rose foliage

Multiflora rose as a spontaneous ornamental in Watertown, Massachusetts

Multiflora rose flowers

Multiflora rose fruits

Rubus allegheniensis Porter   Common Blackberry

Synonyms: *Rubus villosus*, bramble

Life Form: deciduous shrub; 3–8 feet (1–2.5 m) tall

Place of Origin: eastern North America

Vegetative Characteristics: Common blackberry produces erect, green or dark red canes that are covered with stout, sharp spines that can inflict a painful prick on the unwary; individual canes typically live for 2 years before fruiting and dying. Palmately compound leaves have 5 dark green leaflets, red petioles, and typically turn dark burgundy red in fall. The terminal leaflet is always larger than the lateral leaflets and has the longest petiole.

Flowers and Fruit: Clusters of showy white, insect-pollinated flowers—approximately an inch (2.5 cm) wide—terminate 2-year-old canes in late May and June. These are followed by sweet, black berries in July. Blackberries have a solid core (*receptacle*) when picked, as opposed to raspberries, which are hollow because the receptacle remains attached to the plant.

Germination and Regeneration: Numerous animals relish blackberries and disperse the seeds that germinate in a variety of habitats from full sun to partial shade. New shoots can arise either adjacent to the old canes from perennial crowns or some distance away on deep, underground stems.

Habitat Preferences: Common blackberry tolerates a wide variety of soil and light conditions but grows best in deep, moist soil and full sun. In the countryside, it grows in old fields, thickets, and woodland edges. In the urban environment, it is common in minimally maintained public parks, vacant lots, unmowed meadows and fields, woodland edges, and dry, sunny slopes.

Cultural Significance: The fruit, which is often produced in abundance, is prized for making pies, jams, and jellies. In traditional medicine, a tea made from the root has been used to treat diarrhea and stomach pain and as a "female tonic."

Related Species: Cutleaf or **evergreen blackberry** (***Rubus laciniatus* (Weston) Willd.**) is native to Eurasia and produces palmately compound leaves with 5 deeply toothed leaflets, each of which can be further divided into smaller sub-leaflets. Stems (and leaf petioles) are covered with vicious, curved thorns and, in good soil, the plant can grow up to 10 feet (3 m) tall. In areas with mild winters, this species is evergreen. The pink to white flowers have 5 petals that are typically cut into 3 lobes at their tips. Its shiny, black fruits are similar to those of the common blackberry and are readily consumed by both people and animals. The plant grows well along disturbed forest edges, thickets, and meadows and is considered invasive in many parts of the country; it is especially rampant in the Pacific Northwest.

A vigorous stand of young blackberry shoots

Blackberry foliage

Blackberry bushes in full flower

Blackberry flowers

Cutleaf blackberry leaf

Blackberry fruits

99

Rubus flagellaris Willd.   Northern Dewberry

Synonyms: *Rubus canadensis*, whip dewberry

Life Form: prostrate, deciduous vine; stems up to 2 feet (60 cm) tall by 10 feet (3 m) long

Place of Origin: eastern North America

Vegetative Characteristics: Dewberry is a completely prostrate vine with sharp, backward-curving prickles scattered along its creeping stems. Compound leaves typically have 3 leaflets (occasionally 5) that are 0.75–1.5 inches (1.8–3.7 cm) long with toothed margins.

Flowers and Fruit: Like the blackberry, northern dewberry produces large, white, insect-pollinated flowers—approximately 1 inch (2.5 cm) wide—in May and June, followed by reddish-black fruit about 0.5 inch (1.2 cm) wide.

Germination and Regeneration: Dewberry fruits are widely dispersed by birds and small rodents, and the seeds germinate in a variety of disturbed habitats. New stems emerge from the perennial woody crown at the base of the plant, and the tips of the trailing stems root where they touch the ground.

Habitat Preferences: Dewberry is common in sunny, dry sites with sandy or compacted soil, including urban meadows and minimally maintained grasslands. People who walk through such habitats often become entangled in the plants. In the immortal words of Darlington and Thurber (1859, 127): "There is scarcely a farmer's boy who is not well acquainted with it, from having often encountered its prickly trailing stems with his naked ankles, while heedlessly traversing the old fields where it abounds."

Environmental Functions: Food and habitat for wildlife; erosion control on slopes.

Cultural Significance: Northern dewberries are considered to be the earliest and sweetest of our native blackberries and excellent for preserves and pies. An extract of the root, like that of the blackberry, was once used to treat diarrhea and stomach pain and as a "female tonic."

Foliage of northern dewberry

Prostrate stems of northern dewberry

Northern dewberry flowers

Northern dewberry fruits

Northern dewberry stems trailing over an embankment

Rubus occidentalis L.   Black Raspberry

Synonyms: blackcap, thimbleberry

Life form: small deciduous shrub; 4–6 feet (1.5–2 m) tall

Place of Origin: eastern North America

Vegetative Characteristics: Black raspberry produces alternate, compound leaves, usually with 3 (sometimes 5) leaflets that are 2–4 inches (5–10 cm) long, and are pale green above and whitish below. During the growing season, its stems are conspicuously covered with a chalky, bluish-white powder and are armed with scattered, hooked prickles; in winter the stems turn a distinctive purplish maroon.

Flowers and Fruit: Two-year-old black raspberry stems produce terminal clusters of small-petaled white flowers, approximately 0.5 inch (1.2 cm) wide, in early summer. Following insect pollination, delicious purple-black fruits follow in mid-to-late summer; the canes die after they produce fruit. The core of the fruit (the receptacle) remains behind when the berry is picked, which distinguishes it from the blackberry receptacle that detaches with the fruit.

Germination and Regeneration: Fruits are dispersed by birds and other animals, and seeds germinate in sun or shade; new stems arise from a perennial base; 2-year-old stems arch over and root at the tip, giving rise to new plantlets.

Habitat Preferences: Black raspberry is an adaptable species that can tolerate a wide variety of soil conditions in either sun or shade. In the urban environment, it is common along the margins of fields or woodlands, in vacant lots and waste dumps, in the understory of open woodlands, and on dry slopes.

Cultural Significance: The purple black fruits are excellent for eating right off the bush and for making pies and jam. A tea made from the root has been used medicinally to treat diarrhea and stomach pain and as a "female tonic."

Related Species: Wineberry (*Rubus phoenicolasius* Maxim.) is an Asian raspberry that was introduced into North America in 1890 for its edible fruits. It can grow to approximately 6 feet (1.9 m) tall with arching stems that root at their tips. It produces small, white, insect-pollinated flowers and compound leaves that are composed of 3 leaflets; the terminal leaflet, which is about 4 inches (10 cm) long, is larger than the 2 lateral leaflets. The plant is easily recognizable because all of its parts are densely covered with reddish glandular hairs that glisten in the sunlight. It produces copious amounts of delicious red fruits in July that are widely dispersed by birds. This species grows vigorously in moist, sunny sites; several states list it as an invasive species.

Black raspberry stems are chalky white with scattered thorns

Black raspberry leaves typically have 3 leaflets

Black raspberry flower

Black raspberry fruits

Wineberry fruits

Wineberry stems are covered with soft, red spines

Wineberry foliage and fruits

Phellodendron amurense Rupr.   Amur Corktree

SYNONYMS: *P. japonicum, P. sachalinense, P. lavallei*

LIFE FORM: **deciduous tree**; up to 60 feet (18 m) tall

PLACE OF ORIGIN: northeast Asia

VEGETATIVE CHARACTERISTICS: Amur corktree is a fast-growing species with wide-spreading branches, a flat-topped crown, and corky, deeply ridged bark. Opposite, pinnately compound leaves are dark green, glossy, and smooth. They are roughly a foot (30 cm) long, have an odd number of leaflets (from 5–13), and turn bright yellow in early fall. Bright yellow roots are conspicuous on seedlings pulled out of the ground.

FLOWERS AND FRUIT: Amur corktree produces greenish-yellow, insect-pollinated flowers in late spring on separate male and female individuals. These are followed by clusters of oily black fruits, some of which remain on the tree through the winter. European starlings and mourning doves feed on—and disperse—the seeds after they fall to the ground in early spring.

GERMINATION AND REGENERATION: Seeds germinate readily on disturbed, open ground as well as in the shady forest understory. Damaged trees sprout readily from the base or from the roots.

HABITAT PREFERENCES: Amur corktree tolerates both shade and drought as well as a variety of soil conditions ranging from acid to alkaline; it is becoming an increasingly common component of disturbed or emergent woodlands adjacent to areas where it has been planted. It can also be found on rock outcrops, at the base of stone walls, and in neglected ornamental planting beds. While not yet particularly common in urban areas, this could change in the not-too-distant future.

ECOLOGICAL FUNCTIONS: Heat reduction in paved areas; tolerant of roadway salt and compacted soil; food and habitat for wildlife; erosion control on slopes.

CULTURAL SIGNIFICANCE: Amur corktree is native to northeastern Asia, most notably the watershed of the Amur River, which separates China from Russia. It was introduced into North America from St. Petersburg, Russia, in 1874 by the Arnold Arboretum in Boston, which began distributing seedlings in the early 1900s. With the help of birds, the tree has spread from established plantings into surrounding woodlands, and several states now list it as an invasive species. The yellow roots of Amur corktree contain the alkaloid compound *berberine* and are used in traditional Chinese medicine as a digestive tonic and a purgative.

Mature Amur corktree with flat-topped growth habit

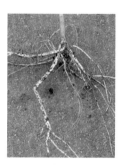

Amur corktree roots

Amur corktree foliage

Amur corktree bark

Amur corktree fall foliage and fruit

Amur corktree fruits can stay on female trees all winter

Populus deltoides Marshall   Eastern Cottonwood

SYNONYMS: eastern poplar, necklace poplar

LIFE FORM: **deciduous tree**; 50–80 feet (15–25 m) tall

PLACE OF ORIGIN: eastern North America

VEGETATIVE CHARACTERISTICS: Bark is ash-gray to brown and, on old trees, is corrugated with deep ridges and furrows. Eastern cottonwood is a fast-growing species, and young trees typically display a leggy, sparsely branched form with smooth, greenish-yellow stems. Young twigs are reddish and often have prominent corky "wings." Simple, alternate leaf blades are roughly triangular (deltoid), 3–5 inches (7.5–12.5 cm) long, and have serrated margins. The flattened petioles, which are often red, catch the slightest breeze, causing the leaves to quiver. Leaves on vigorous young sprouts often develop a vertical rather than horizontal orientation and are typically much larger than those found on the branches of older trees. It leafs out in early spring and its fall color is pale yellow.

FLOWERS AND FRUIT: Like all poplars, eastern cottonwood is dioecious, with separate male and female individuals. Wind-pollinated flowers are produced in early spring, well before the leaves come out. The pendulous male catkins are bright red and several inches (up to 7.5 cm) long; female flowers are small and inconspicuous. In late spring, the female capsules release thousands of tiny, cottony white seeds that accumulate in large masses near the base of the tree and are the source of its common name. Some people claim to be allergic to cottonwood "fluff," but they are most likely reacting to the inconspicuous grass pollen that is abundant at the same time.

GERMINATION AND REGENERATION: Cottonwood's wind-dispersed seeds germinate without a dormancy delay in late spring. Seedlings grow rapidly on bare soil in full sun. Open-grown trees typically have multiple trunks, but it does not root sucker or form large clones like quaking aspen.

HABITAT PREFERENCES: In native habitats, eastern cottonwood is a moisture-loving species that grows along the margins of freshwater wetlands, ponds, streams, and rivers. In urban areas, it grows in disturbed wetlands and drainage channels, in vacant lots and rubble dumps, and along railroad tracks with ballast substrate.

ECOLOGICAL FUNCTIONS: Heat reduction in paved areas; stream and riverbank stabilization; food and habitat for wildlife.

CULTURAL SIGNIFICANCE: The inner bark contains an aspirin-like compound (*salicin*), and Native Americans used it as a tea to treat scurvy and as a "female tonic." Eastern cottonwood has hybridized with *Populus nigra* from Europe to produce the fast-growing *P.* x *canadensis* (Carolina poplar), which is widely cultivated for landscape and forestry purposes.

Mature eastern cottonwood in Boston

Eastern cottonwood resprouting following injury

Eastern cottonwood leaf in autumn, about 3 inches (7 cm) wide

Eastern cottonwood bark

Eastern cottonwood male catkins about to shed their pollen in early spring

Freshly dispersed eastern cottonwood seeds

Populus tremuloides Michx.   Quaking Aspen

SYNONYMS: trembling aspen, American aspen, popple, chatterbox tree

LIFE FORM: **deciduous tree**; up to 50 feet (15 m) tall

PLACE OF ORIGIN: northern, central, and eastern North America

VEGETATIVE CHARACTERISTICS: Quaking aspen is a fast-growing, upright tree. The bark of young trees is smooth and yellow-green with black markings; mature trees have dark brown, furrowed bark. Leathery leaf blades are round or egg-shaped with coarse teeth and are 2–4 inches (5–10 cm) long and wide. The distinctly flattened petioles cause the leaves to "tremble" in the slightest breeze. It produces leaves in early spring, which turn bright yellow in the fall.

FLOWERS AND FRUIT: Male trees produce catkins that are 2–3 inches (5–7.5 cm) long and release their wind-dispersed pollen in early spring before the leaves come out. Female trees shed their tiny "cottony" seeds in late spring.

GERMINATION AND REGENERATION: Seeds germinate shortly after dispersal, on bare soil in full sun. Established trees reproduce vigorously from root suckers following injury, eventually forming large clumps of stems. In the West, quaking aspen groves consisting of thousands of trunks, all with the same genetic makeup, cover hundreds of acres and are thousands of years old.

HABITAT PREFERENCES: Quaking aspen is one of the most widespread trees in North America; it grows best in sunny, dry sites where competition from other trees is reduced. In the urban environment, it is common on abandoned rubble dumps, in vacant lots, in open woods and thickets, and along railroad tracks and unmowed highway banks. In rural areas, it is a common colonizer of disturbed, sandy sites.

CULTURAL SIGNIFICANCE: Quaking aspen bark contains the aspirin-like compound *salicin*, an anti-inflammatory agent, analgesic, and fever reducer. Native Americans made tea from the bark to treat excessive menstrual bleeding, stomach pain, venereal disease, urinary aliments, worms, colds, and fevers. Today the tree is an important source of pulpwood for book and magazine paper.

RELATED SPECIES: **White** or **silver poplar (*Populus alba* L.)** is native to Europe. It produces alternate, simple leaves with 3–5 coarse lobes and distinctive fuzzy white undersides. Leaves are 2–4 inches (5–10 cm) long and nearly as broad, and they turn yellow in the fall. This species was widely planted in the mid-1800s through the early 1900s as a windbreak but is not planted today. It exists mainly as a spontaneous plant reproducing from root suckers in neglected urban woodlands and along the margins of rivers and streams. Darlington and Thurber (1859, 332) reported that "some of the grass-plots in the public squares of New York have been quite overrun by the side-spreading suckers of this tree; even in closely-paved streets they work their way up between the stones. It should be discarded altogether." The Greek physician Dioscorides included white poplar in his first-century herbal, *De Materia Medica.*

Mature stand of quaking aspen

Quaking aspen in the ballast along a Detroit rail line

Quaking aspen foliage

Clonal stand of quaking aspen along the Massachusetts Turnpike

White poplar trunk and branch

White poplar foliage; note the hairy white underside

SPECIES: Several native shrubby willows—including **Salix bebbiana Sarg., S. discolor Muhl.,** and **S. humilis Marshall**—are common in the Northeast and difficult to tell apart. As Henry David Thoreau noted in his *Journal,* "The more I study willows, the more I am confused." Two shrubby willows introduced from Europe are also common: the purple osier (**S. purpurea** L.) and the osier or basket willow (**S. viminalis** L.).

LIFE FORM: **multistemmed shrubs**; 10–20 feet (3–6 m) tall

PLACE OF ORIGIN: eastern North America and Eurasia

VEGETATIVE CHARACTERISTICS: Shrubby willows typically have a multistemmed habit; the smooth, green, red, or yellow stems turn brown with age. Hairless leaves are typically oval to narrowly lanceolate and often have whitish or silvery undersides. The single, smooth bud scale identifies winter buds. Leaves typically turn a nondescript yellow in the fall.

FLOWERS AND FRUIT: Shrubby willows are dioecious and their flowers are typically pollinated by insects but also by wind in some cases. The male flowers (catkins) are showier than those of the females and are responsible for the early spring display of pussy willows. Fruit capsules on female trees mature in late spring, releasing tiny seeds topped with silky hairs that facilitate their dispersal by wind.

GERMINATION AND REGENERATION: Willow seeds germinate in moist, sunny sites immediately after being shed—they have no dormancy period. Established plants produce new stems from the base, and leaning stems will produce roots where they come in contact with the soil.

HABITAT PREFERENCES: Shrubby willows grow best along the sunny margins of freshwater wetlands, ponds, streams, and rivers and in roadside drainage ditches.

ECOLOGICAL FUNCTIONS: Nutrient absorption in wetlands; tolerant of roadway salt; food and habitat for wildlife; stream and riverbank stabilization.

CULTURAL SIGNIFICANCE: Basket and purple osier willows were introduced into North America from Europe as a source of supple stems used in the production of containers and wattle fences. The bark of most willows contains salicin, a precursor of commercial aspirin, and it is still used in European traditional medicine. Various species of shrubby willow have been used for phytoremediation on sites contaminated with heavy metals (especially cadmium); for biomass production in short-rotation forestry; and for erosion control along stream banks.

RELATED SPECIES: The familiar **weeping willow** (*Salix* x *sepulcralis* **Simonk.** [formerly *S. alba* 'Tristis']) is a large tree with strongly weeping branches and yellow twigs in spring that has been widely planted throughout the Northeast as an ornamental along the banks of rivers, streams, and ponds. It can propagate itself when twigs or branches break off in a storm, float downstream, and take root along the sandy shoreline. Large trees that blow over often re-root and continue growing.

Shrubby willow growing in a moist microhabitat on the second story of an abandoned Detroit factory

Typical foliage of a shrubby willow with sharp-pointed lateral buds

Shrubby willow growing in a sunny wetland

Male flowers of a shrubby willow

Shrubby willow growing out of a canal wall in Cambridge, Massachusetts

Acer negundo L.   Box Elder

Synonyms: ash-leaved maple, Manitoba maple, maple ash

Life Form: deciduous tree; up to 60 feet (18 m) tall

Place of Origin: North America

Vegetative Characteristics: Box elder typically produces a short, leaning trunk or multiple trunks with a broad, spreading crown. Opposite, pinnately compound leaves consist of 3–7 coarsely toothed leaflets, 2–4 inches (5–10 cm) long, that are extremely variable in shape. Young twigs are purplish or green and are often covered with a white, waxy coating that rubs off easily. Leaves turn pale yellow in the fall.

Flowers and Fruit: Box elder produces conspicuous chains (racemes) of green, wind-pollinated flowers, 4–6 inches (10–15 cm) long, in late spring. These are followed by strings of winged samaras that persist through most of the winter, making the tree easy to identify in the leafless condition. Box elder produces unisexual flowers on separate male and female (seed-bearing) trees.

Germination and Regeneration: Seeds are wind dispersed from fall through winter and germinate on moist, bare ground in spring. Saplings and mature trees sprout readily from the base following injury or partial uprooting of the trunk, producing a multistemmed growth form.

Habitat Preferences: This disturbance-adapted, native species typically grows along streams and rivers in full sun. In the urban environment, it grows in a wide variety of sites including the margins of freshwater wetlands, vacant lots, railroad rights-of-way, small pavement openings, rock outcrops, and along chain-link fence lines. It is a remarkably hardy and adaptable tree with a widespread distribution across most of Canada and the United States, and even into Mexico and Guatemala.

Ecological Functions: Food and habitat for wildlife; stream and riverbank stabilization; soil building on degraded land; heat reduction in paved areas; tolerant of roadway salt and compacted soil.

Cultural Significance: Native Americans made a tea from the bark to induce vomiting, and the sap in spring can be boiled down to make syrup. Horticultural varieties of box elder with variegated foliage or red fall color have limited availability in the nursery trade. The common name of the plant derives from the resemblance of its wood to that of boxwood (genus *Buxus*) and of its foliage to that of the elderberry (genus *Sambucus*). It was introduced into Europe at an early date and is now considered an invasive species in many countries. The tree is host to the box elder beetle (*Boisea trivittata*), a bug approximately 0.5 inch (1.2 cm) long with red wing markings. They feed on the seeds produced by female plants but do not damage the trees themselves. In areas where the tree is common, the bugs are an annoyance to homeowners because they smell bad and congregate in large numbers on the sides of houses.

Box elder
growth habit

Box elder seedling with juvenile foliage

Compound leaves of box elder

Multistemmed growth form of a
cut-back plant, note the silvery
white stems

Box elder seeds

Box elder in
flower (male)

Acer platanoides L.   Norway Maple

SYNONYMS: none

LIFE FORM: **deciduous tree**; 40–60 feet (12–18 m) tall

PLACE OF ORIGIN: northern and central Europe and western Asia

VEGETATIVE CHARACTERISTICS: Norway maple develops a broad, spreading crown; its bark is smooth and gray when young and becomes furrowed with age. Opposite, dark green leaves are 4–7 inches (10–17.5 cm) wide and have the classic maple leaf shape—with 5 primary lobes. They turn yellow in late fall, long after the native maples have dropped their leaves, and the petiole exudes milky sap when broken. The tenacious roots of Norway maple are very shallow and typically keep other plants from growing beneath it.

FLOWERS AND FRUIT: Norway maple produces clusters of distinctive chartreuse, wind-pollinated flowers in early spring before the leaves emerge; these are followed by paired samaras with wings that spread apart at an angle of about 120°. The head of the seed containing the embryo is relatively flat compared with that of other maples.

GERMINATION AND REGENERATION: The wind-dispersed seeds mature in the fall and can travel 100 feet (30 m) or more from their parent tree. Seeds will germinate in either sun or shade, and saplings resprout readily from the base following damage.

HABITAT PREFERENCES: Because it tolerates shade and compacted soils, Norway maple has become a dominant late successional tree in many urban and suburban woodlands. It is also common along streams and riverbanks, in neglected residential and commercial landscapes, in vacant lots and dumps, and along chain-link fence lines.

ECOLOGICAL FUNCTIONS: Tolerant of roadway salt and compacted soil; heat reduction in paved areas; erosion control on slopes; food and habitat for wildlife.

CULTURAL SIGNIFICANCE: Norway maple is among the most commonly planted street trees in the Northeast. Numerous horticultural selections are available, including purple-leaved ('Crimson King') and fastigiated cultivars. Norway maple was first imported into North America in 1756 by John Bartram of Philadelphia and became popular during the late 1800s. It was extensively planted from the 1940s through the 1970s to replace American elms wiped out by Dutch elm disease. Many states now list it as an invasive species.

RELATED SPECIES: **Sugar maple (*Acer saccharum* Marshall)** is native to eastern North America and does not grow as well in cities as Norway maple does. The two species can be distinguished by the fact that sugar maple petioles do not exude a milky sap when broken, and its fall color is red to orange rather than yellow.

Butter yellow Norway maple leaves in late fall

Chartreuse Norway maple flowers in early spring

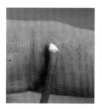

Milky sap distinguishes Norway maple from other maples in this book

Developing seeds of Norway maple

Typical Norway maple leaf

Mature Norway maple bark

Acer pseudoplatanus L.   Sycamore Maple

SYNONYM: planetree maple

LIFE FORM: **deciduous tree**; up to 80 feet (25 m) tall

PLACE OF ORIGIN: Europe and western Asia

VEGETATIVE CHARACTERISTICS: Sycamore maple produces coarsely toothed, op-posite leaves that are 3–6 inches (7.5–15 cm) long and wide, often with red petioles; the smooth, 5-lobed leaves have a leathery texture and are dark green above and greenish-white beneath. Bark on mature trees can vary from pale gray to reddish brown, and it flakes off in large, irregular patches. Leaves turn pale yellow in fall.

FLOWERS AND FRUIT: Sycamore maple produces small, yellow-green, wind-pollinated flowers in May on pendulous chains (racemes) that are 2–6 inches (5–15 cm) long. Female flowers are usually at the top and male flowers on the bottom. These are followed by chains of winged seeds (samaras) that mature in the fall. Seeds are produced in pairs and have broad wings that form a 60–90° angle.

GERMINATION AND REGENERATION: Sycamore maple seeds are dispersed by wind in autumn, and seedlings germinate in spring under a wide variety of conditions from full sun to partial shade. The tree is quite tolerant of salt spray and dry soil.

HABITAT PREFERENCES: Because of its salt tolerance, sycamore maple was widely planted as an ornamental in coastal landscapes up and down the East Coast. It has now fallen out of favor with landscapers, but mature specimens still persist in the landscape and seedlings grow well in a variety of disturbed urban sites and along roadways treated with salt.

ECOLOGICAL FUNCTIONS: Tolerant of salt along roadways and near the ocean; heat reduction in paved areas; erosion control on slopes; tolerant of compacted soil; food and habitat for wildlife.

CULTURAL SIGNIFICANCE: Sycamore maple was probably introduced into North America from Europe during the early 1800s; it was widely planted as a street or park tree during the late 1800s and the first half of the twentieth century. Horticultural selections of this species are numerous, most with brightly colored or variegated foliage.

Sycamore maple foliage and seeds

Sycamore maple growth habit

Mature bark of sycamore maple

Leaves of a vigorous sycamore maple sprout

Sycamore maple seeds

Sycamore maple inflorescence

Acer saccharinum L. Silver Maple

SYNONYMS: *Acer dasycarpum*, soft maple

LIFE FORM: deciduous tree; up to 100 feet (30 m) tall

PLACE OF ORIGIN: eastern North America

VEGETATIVE CHARACTERISTICS: Silver maple is one of the tallest trees native to the Northeast. It has light gray bark and a broad, open crown, and it often produces multiple trunks. Opposite leaves are approximately 6 inches (15 cm) long and typically have 5 lobes; the margin of the leaf is coarsely serrated and the sinuses separating the lobes extend nearly to the central vein. The upper surface of the leaf is light green and the underside silvery white; fall color is pale yellow.

FLOWERS AND FRUIT: While not strictly a dioecious species, individual silver maple trees (as well as most other maples) produce flowers that are predominantly male or female, so it usually takes more than one tree to produce viable seeds. Silver maple is among the first of our native trees to bloom, producing small, reddish, wind-pollinated flowers from mid-March through early April. Female flowers quickly mature into broad-winged samaras up to 2 inches (5 cm) long.

GERMINATION AND REGENERATION: Seeds are dispersed by the wind in late spring, before the leaves have fully expanded; they germinate without delay on exposed, sunny sites. Established trees sprout readily from the base when injured or partially uprooted by stream flow, resulting in a multistemmed growth form.

HABITAT PREFERENCES: Silver maple grows best in sunny, moist habitats, especially freshwater wetlands and along the banks of ponds, streams, and rivers.

ECOLOGICAL FUNCTIONS: Heat reduction in paved areas; food and habitat for wildlife; stream and riverbank stabilization; nutrient absorption in wetlands; tolerant of salt and compacted soil.

CULTURAL SIGNIFICANCE: Silver maples were once widely planted as street trees because of their rapid growth, but this has led to problems because they often interfere with overhead wires and generate cracks in sidewalks. The species also has a reputation for being weak wooded and dropping branches, but this is usually the result of poor pruning done to compensate for planting the tree underneath utility wires or in locations that are too small for its large size.

RELATED SPECIES: Red maple (*Acer rubrum* L.) is native to eastern North America and can reach heights up to 100 feet (30 m). Although red maple can grow in a variety of habitats, it is most common in bottomland situations or sites with good soil moisture. It produces bright red flowers in very early spring on separate male and female plants and disperses its relatively small—about 0.75 inch (1.8 cm) long—winged seeds in late spring. Leaves are opposite and typically have 3, but often 5, lobes with serrated margins; they turn bright red or yellow in the fall. The bark of young trees is smooth and light gray, but it becomes rough and scaly with age. Horticultural selections of red maple with bright red fall foliage and upright growth habits are widely cultivated as street trees.

A mature silver maple growing along the Charles River in Cambridge, Massachusetts

Silver maple foliage has silvery-white undersides

Female flowers of silver maple in early spring

Silver maple seeds (*right*) and red maple seeds (*left*) are shed in mid- to late spring

Red maple leaves are bright red in fall

Red maple seeds begin to develop before the leaves come out in spring

Ailanthus altissima (P. Mill.) Swingle   Tree-of-heaven

SYNONYMS: *Ailanthus glandulosa*, copal tree, ghetto palm, stink tree, Chinese sumac

LIFE FORM: deciduous tree; up to 70 feet (21 m) tall

PLACE OF ORIGIN: central and northeast China

VEGETATIVE CHARACTERISTICS: The smooth, gray trunk is sparsely branched. Leaves are alternate, pinnately compound, can reach 3 feet (1 m) in length, and are composed of 11–30+ sharp-pointed leaflets 2–6 inches (5–15 cm) long. The leaves, which have an unpleasant odor when crushed, appear late in spring; the leaflets drop with little or no fall color after the first hard frost, leaving the bare leaf-stalk (*rachis*) behind. Stout, sparsely branched twigs with their prominent, heart-shaped leaf scars present a stark appearance in winter. The wood is light and brittle.

FLOWERS AND FRUIT: *Ailanthus* produces small, greenish-yellow flowers in June on separate male and female trees (*dioecious*). Male flowers have a noticeably unpleasant smell that attracts insects to carry the pollen to female trees. Seeds contain an embryo that is located in the center of a slightly twisted, single-winged seed (*samara*), and clusters of these wind-dispersed seeds persist on the tree into winter, making identification easy.

GERMINATION AND REGENERATION: Seeds germinate readily on bare ground in sun or shade. Roots of mature trees can produce numerous suckers, especially after the primary trunk has been damaged or cut down; the stumps of young trees also sprout prolifically. In China, the term "good for nothing *Ailanthus* stump-sprout" refers to a spoiled or irresponsible child who fails to live up to parental expectations (Hu 1979).

HABITAT PREFERENCES: Tree-of-heaven grows well in full sun on dry, rocky, or sandy soil. The tree is famous for its ability to tolerate disturbance and forms dense thickets of stems along unmowed railroad and highway embankments. It seems to have a special affinity for the cracks that form where concrete and blacktop come together. The foliage is reported to contain allelopathic chemicals that suppress the growth of nearby plants.

CULTURAL SIGNIFICANCE: *Ailanthus* bark and leaves have a variety of uses in traditional Chinese medicine, including the treatment of diarrhea, dysentery, and tapeworm. It was introduced into North America as an ornamental in the early 1800s and by 1859, Darlington and Thurber noted that "there is no tree so generally employed in the city of New York as a shade tree" (p. 76). By the early 1900s, *Ailanthus* had become thoroughly naturalized throughout that city—as well as every other city in the Northeast—and was famously celebrated in Betty Smith's 1943 novel, *A Tree Grows in Brooklyn*. *Ailanthus* grows spontaneously in every state in the contiguous U.S. except for the upper Great Plains. In 1996, a fatal fungal wilt disease, *Verticillium albo-atrum*, was discovered on stands of *Ailanthus* in Pennsylvania and New York City.

Mature growth habit of a male tree-of-heaven

Tree-of-heaven growing through a chain-link fence

Tree-of-heaven leafs out in late spring

Tree-of-heaven compound leaves can be more than 3 feet (1 m) long

Seeds of the red-fruited tree-of-heaven, variety *erythrocarpum*

Solanum dulcamara L.   Bittersweet Nightshade

SYNONYMS: bittersweet, bitter nightshade, blue nightshade, poison berry, climbing nightshade, blue bindweed, violet bloom

LIFE FORM: trailing, semi-woody vine; up to 10 feet (3 m) long

PLACE OF ORIGIN: Eurasia and North Africa

VEGETATIVE CHARACTERISTICS: Flexible green or purple stems of bittersweet nightshade trail along the ground or twine around and through other plants. Alternate, dark green leaves are 2–5 inches (5–12.5 cm) long and have 2 distinct forms: one is compound with 3 leaflets, of which the terminal segment is always the largest, and the other is simple and ovate to oval. Foliage has an unpleasant odor when crushed, and plants growing in exposed, sunny sites often take on a dramatic dark purple hue. The ropy roots are white and typically grow just below the soil surface.

FLOWERS AND FRUIT: Bittersweet nightshade produces clusters of small, star-shaped flowers along the stem, each with 5 purple to violet, downward-pointing (reflexed) petals and a cone of yellow anthers in the center. The contrasting colors of the flowers are attractive to pollinating bees. Mature fruit is an oval red berry approximately 0.5 inch (1.2 cm) long that contains numerous small, white seeds. Flowers and fruit are produced throughout the summer.

GERMINATION AND REGENERATION: Fruits are dispersed by birds, and seeds germinate underneath roosts. Prostrate stems root where they touch the ground, and the shallow root system will produce suckers following traumatic injury to the main stem.

HABITAT PREFERENCES: Bittersweet nightshade is a drought-tolerant species that grows best in full sun and rich soil. In the urban environment, it is common in vacant lots and rubble dumps; along chain-link fences, rock outcrops, and stone walls; and growing up through shrubs in ornamental landscapes. In its native habitat, it grows along the banks of rivers, streams, and lakes.

ECOLOGICAL FUNCTIONS: Tolerant of compacted soils; food and habitat for wildlife.

CULTURAL SIGNIFICANCE: As is true of many members of the Solanaceae, the foliage of bittersweet nightshade is toxic and somewhat narcotic, although the ripe red fruits are harmless. The plant has a long tradition of use in European folk medicine to treat warts and skin problems, such as eczema, as well as jaundice and rheumatism. Darlington and Thurber (1859, 252–53) noted that bittersweet nightshade was "extensively naturalized in fertile soils, and is often tolerated and even sometimes cultivated to train over walls and fences, as its flowers and fruit are showy."

Bittersweet nightshade flowers and foliage often have a purplish cast

Bittersweet nightshade growing on a chain-link fence in Boston

Bittersweet nightshade foliage

Bittersweet nightshade flowers

Bittersweet nightshade fruits

Ulmus americana L.   *American Elm*

SYNONYMS: white elm, water elm, soft elm

LIFE FORM: **deciduous tree**; up to 100 feet (30 m) tall

PLACE OF ORIGIN: eastern North America

VEGETATIVE CHARACTERISTICS: American elm has a distinctive upwardly arching, vase-shaped form; older trees develop a broad, spreading crown. Bark is dark gray with broad, deep ridges. Leaves are alternate, oblong to egg-shaped, and 3–6 inches (7.5–15 cm) long by 1–3 inches (2.5–7.5 cm) wide. They have doubly serrated margins, are rough to the touch, and have unequal bases, which gives them a somewhat lopsided appearance. Leaves have short petioles and turn yellow in the fall.

FLOWERS AND FRUIT: American elm produces wind-pollinated, green to yellow flowers in very early spring before the leaves come out. These are followed quickly by conspicuous clusters of papery "wafers"—the seeds—that are dispersed by wind in early June.

GERMINATION AND REGENERATION: Seeds are shed in late spring and germinate immediately on bare mineral soil; they are also shade tolerant and seedlings can often be seen emerging from under shrubs and hedges. Saplings and young trees sprout readily from the base following damage.

HABITAT PREFERENCES: As a wild plant, American elm is a bottomland species that grows best in moist, sunny habitats and along the edges of fields and woodlands. In the urban environment, it grows on moist or dry compacted soils and is common in minimally maintained open space; vacant lots and waste dumps; emergent woodlands; along the margins of freshwater wetlands, ponds, streams, and rivers; chain-link fence lines; and roadside drainage ditches.

ECOLOGICAL FUNCTIONS: Heat reduction in paved areas; tolerant of roadway salt and compacted soil; erosion control on slopes; stream and riverbank stabilization; food and habitat for wildlife.

CULTURAL SIGNIFICANCE: This widespread species has a long history of cultivation in eastern North America because of its beautiful, vase-shaped form and because it is easy to transplant and tolerates heavy pruning. Its wood has a twisted grain and is difficult to split for firewood or to make into lumber. Prior to the arrival of Dutch elm disease in the 1930s, American elms accounted for roughly 50% of the street trees in the Northeast and formed a literal monoculture in many New England towns. Mature specimens are now rare in the landscape, but young trees still grow along rivers and streams and often survive long enough to produce seed, and are fairly common in the wild, albeit with a much shorter lifespan than before the blight. Several cultivars of American elm resistant to Dutch elm disease are available in nurseries.

Serrated leaf of American elm

Growth habit of mature American elm

Developing American elm fruits

Two American elms in Hartford, Connecticut, one pruned brutally, the other allowed to grow freely

American elm fall color

Seedling American elms showing alternate leaf arrangement

Ulmus pumila L.   Siberian Elm

SYNONYMS: littleleaf elm, Chinese elm, Asiatic elm

LIFE FORM: **deciduous tree;** up to 80 feet (24 m) tall

PLACE OF ORIGIN: central and northeast Asia

VEGETATIVE CHARACTERISTICS: Siberian elm has a stiffer, more upright growth habit that does American elm, and its deciduous to semi-evergreen, elliptical leaves are much smaller, approximately 2.5 inches (6.2 cm) long and half as wide. They are arranged alternately along the stem and have coarsely serrated margins and are only slightly asymmetrical at the base. The tree holds its leaves until late into the fall (November) before they turn yellow brown and fall off. With age, the tree produces a thick trunk covered with dark brown, deeply fissured bark.

FLOWERS AND FRUIT: Siberian elm produces wind-pollinated, greenish flowers in early spring before the leaves come out. These are followed quickly by clusters of thin, oval seeds that are approximately 0.5 inch (1.2 cm) across with a distinct notch at their tip; they are dispersed by the wind in early June.

GERMINATION AND REGENERATION: Seeds germinate in spring shortly after being shed from the tree; mature trees can reproduce vigorously from root suckers following damage to the main trunk.

HABITAT PREFERENCES: Siberian elm grows in a variety of disturbed, sunny habitats including abandoned railroad tracks, vacant lots, and along stream and riverbanks. Under favorable conditions, it can form dense stands.

ECOLOGICAL FUNCTIONS: Heat reduction in paved areas; tolerant of roadway salt and compacted soil; erosion control on slopes; stream and riverbank stabilization; food and habitat for wildlife.

CULTURAL SIGNIFICANCE: While Siberian elm has been long cultivated in Europe, it does not seem to have gained attention in North America until the early 1900s when hardy strains were introduced from China. It quickly developed a favorable reputation because of its rapid growth, extreme cold hardiness, and drought and wind tolerance—in short, its capacity to grow where few other trees could survive. During the Dust Bowl era of the 1930s, Siberian elm was widely planted as a shelterbelt tree in the upper Midwest and Canada. On the East Coast, Siberian elm was planted as a street tree because of its rapid growth and resistance to Dutch elm disease, but it lost popularity because of its susceptibility to elm leaf beetle, habit of constantly shedding small branches, and tendency to drop large limbs in ice and snow storms.

RELATED SPECIES: **Japanese zelkova (*Zelkova serrata* (Thunb.) Makino)** was widely promoted as a replacement for the American elm in the 1950s and '60s following its elimination from city streets by Dutch elm disease. While not nearly as big as American elm or as graceful, it is a great tree in its own right and deserves to be planted for its disease resistance, elegant form, excellent fall color, and drought tolerance. Zelkova does, however, show a tendency to escape from cultivation, and saplings can often be found growing along shady, woodland edges in the proximity of mature trees.

Remnant Siberian
elm street tree in
Detroit

Mature planting of Siberian
elm in Pennsylvania

Spontaneous Siberian elm in
Detroit

Bonsaied Siberian elm
enmeshed in a chain-link
fence in Boston

American elm (*left*) and
Siberian elm (*right*) growing
side by side in Detroit

Dense stand of
spontaneous
Japanese zelkova
seedlings in
Boston

Ampelopsis glandulosa var. *brevipedunculata* (Maxim.) *Momiy.*   Porcelain Berry

SYNONYMS: *Ampelopsis heterophylla*, Amur peppervine, Asiatic creeper

LIFE FORM: **deciduous vine (liana)**; up to 20 feet (6 m) long

PLACE OF ORIGIN: northeast Asia

VEGETATIVE CHARACTERISTICS: Alternate, light green leaves may be up to 5 inches (12.5 cm) long and have hairy undersides. Leaves on the older parts of the stem (mature leaves) have 3 well-defined lobes; the juvenile leaves produced at the ends of the vigorously growing shoots are deeply dissected and irregularly toothed. The plant climbs by means of tendrils that are produced opposite the leaves; old plants have flexible stems with tight brown bark and prominent lenticels. Leaves typically drop without changing color.

FLOWERS AND FRUIT: Porcelain berry produces inconspicuous clusters of greenish-white, 5-petaled flowers along its stems from July through August. They are insect-pollinated and are followed by hard, round fruits that are initially yellow, green, pink, or lilac and eventually turn into beautiful marbled shades of blue and white (the source of its common name) from September through October.

GERMINATION AND REGENERATION: Mature porcelain berry fruits are readily consumed by birds, which disperse the seeds across the landscape; trailing stems root where they touch the ground, facilitating the rapid spread of this vine.

HABITAT PREFERENCES: Porcelain berry grows in a wide variety of disturbed habitats in full sun to partial shade, but it does best when it has access to good soil moisture. It is particularly common along highway banks but can also be found along forest edges and thickets, rock outcrops and stone walls, and chain-link fences—all habitats where it can overwhelm adjacent vegetation.

ECOLOGICAL FUNCTIONS: Food and habitat for wildlife; erosion control on slopes; stream and riverbank stabilization.

CULTURAL SIGNIFICANCE: Porcelain berry was introduced into North America from Japan in 1875 by Thomas Hogg (Del Tredici 2017a). It quickly became a popular ornamental because of its beautiful blue fruits and finely dissected leaves. It has escaped from cultivation in areas where it was planted and is now listed as an invasive species in many states. The cultivar 'Elegans' has green and white variegated leaves.

Porcelain berry growth habit

Rampant porcelain berry along the Saw Mill River Parkway near New York City

Highly dissected juvenile leaves of porcelain berry

Mature leaves and fruit of porcelain berry

Ripening porcelain berry fruits

Parthenocissus quinquefolia (L.) Planchon   Virginia Creeper

SYNONYMS: *Ampelopsis quinquefolia,* woodbine, five-leaved ivy

LIFE FORM: **deciduous vine (liana)**; up to 60 feet (18 m) long

PLACE OF ORIGIN: eastern North America

Vegetative Features: Alternate, palmately compound leaves typically consist of 5 leaf-lets, 1.5–4 inches (4–10 cm) long, with coarsely toothed margins and a long petiole. The plant climbs by means of branched tendrils with oval disks at their tips that secrete an adhesive substance that binds them tightly to a variety of surfaces—especially brick. Leaves turn brilliant red in the fall.

FLOWERS AND FRUIT: Virginia creeper produces short spikes of inconspicuous greenish-white, insect-pollinated flowers in July that are usually hidden by the leaves. These are followed by small, blue-black berries in September and October.

GERMINATION AND REGENERATION: Fruits are dispersed by birds, and seeds germinate in sun or shade. Stems that contact the ground will form adventitious roots.

HABITAT PREFERENCES: Virginia creeper grows in a wide variety of urban habitats but is most often seen climbing trees along woodland edges; it is also common along dry, sunny roadsides and on stone walls and chain-link fences.

ECOLOGICAL FUNCTIONS: Food and habitat for wildlife; erosion control.

CULTURAL SIGNIFICANCE: Native Americans used this plant for a variety of me-dicinal purposes but mainly as an astringent and a diuretic. Virginia creeper has a long history of cultivation in both North America and Europe for covering walls and buildings because of its spectacular crimson to deep purple fall color.

RELATED SPECIES: **Boston ivy (*Parthenocissus tricuspidata* (Sieb. & Zucc.) Planchon)** is a deciduous, woody vine native to eastern Asia that produces dark green, 3-lobed, entire leaves—4–8 inches (10–20 cm) wide—with a glossy sheen. Typically planted to climb up masonry walls, Boston ivy occasionally spreads beyond the areas where it is planted when bird-dispersed seeds manage to germinate. The foli-age turns a dramatic deep red to maroon in the fall. The plant was first introduced in 1861 from Japan into England and from there into the United States. The term *Ivy League* supposedly originated because the plant was used to cover the brick buildings of exclusive colleges in the Northeast, especially in Boston.

Virginia creeper in brilliant fall color in Boston

Virginia creeper fruits

Virginia creeper climbs by means of tendrils

Virginia creeper leaves

Boston ivy growing under a dry, shady highway

Boston ivy foliage

Vitis riparia Michx. Riverbank Grape

SYNONYMS: frost grape, wild grape

LIFE FORM: deciduous vine (liana); climbing up to 60 feet (18 m)

PLACE OF ORIGIN: eastern North America

VEGETATIVE CHARACTERISTICS: Riverbank grape produces simple, alternate leaves with 3 lobes and relatively few hairs on the underside; they are palmately veined with toothed margins and are 4–8 inches (10–20 cm) long. Vines climb by means of forked, coiling red tendrils that are produced along the stem opposite the leaves. With age, the dark brown stems can become 2–6 inches (5–15 cm) thick with bark shedding off in thin strips. Leaves turn yellow in the fall.

FLOWERS AND FRUIT: Separate male and female plants produce inconspicuous chains (panicles) of greenish-yellow, insect-pollinated flowers in late spring to early summer; they arise from the axils of the leaves and are approximately 4 inches (10 cm) long. The purple-black fruits are smaller than cultivated grapes, about 0.25–0.5 inch (6–12 mm) in diameter, and are produced abundantly in autumn.

GERMINATION AND REGENERATION: Riverbank grapes are eaten by a variety of mammals as well as birds—the seeds germinate freely under their roosts; trailing stems root where they touch the ground.

HABITAT PREFERENCES: Seedlings are highly shade tolerant; once they reach the sunny forest canopy the vines spread out and begin flowering. Wild grapes are vigorous climbers that can easily overwhelm adjacent vegetation. In the urban environment, they are common in the understory of moist woods and thickets, along the banks of streams and rivers, climbing chain-link fences, and along the top of roadside guardrails. Like most vines, wild grapes grow best when their roots are situated in moist, shady soil and their leaves are in full sun.

ECOLOGICAL FUNCTIONS: Tolerant of roadway salt; food and habitat for wildlife; erosion control on slopes.

CULTURAL SIGNIFICANCE: The fruit is edible and, with enough added sugar, makes excellent jelly. Native Americans made a tea from the leaves for medicinal purposes. Established wild grapes can be very destructive to forest trees, particularly when they are entangled with branches that are weighted down by heavy, wet snow or ice.

RELATED SPECIES: The leaves of **fox grape (*Vitis labrusca* L.)** have a dense covering of brownish or whitish hairs on the underside, giving them a rusty or grayish appearance when they blow in the wind. Fruits are larger—about 0.75 inch (1.8 cm) in diameter—and sweeter than riverbank grape. This native woody vine was one of the parents of the famous hybrid 'Concord' grape developed in 1852 by Ephraim Bull of Concord, Massachusetts.

Riverbank grape on a telephone line in Detroit

Riverbank grape overwhelming a pair of arborvitae trees

Riverbank grape fruits

Tangle of riverbank grape stems

Riverbank grape foliage

The woolly undersides of fox grape leaves are distinctive

Amaranthus retroflexus L.   Redroot Pigweed

SYNONYMS: rough pigweed, green amaranth

LIFE FORM: **summer annual**; up to 6 feet (2 m) tall

PLACE OF ORIGIN: Central America

VEGETATIVE CHARACTERISTICS: The upright, hairy stems and stout taproot are typically suffused with red pigment. Simple leaves are alternate, egg- or diamond-shaped, with wavy margins, and can be up to 4 inches (10 cm) long. The uppermost leaves are typically lance-shaped and smaller than the lower ones.

FLOWERS AND FRUIT: Redroot pigweed produces spikes of small, tightly clustered green flowers that form an erect terminal inflorescence. Separate male and female flowers are produced on the same plant and are mainly wind-pollinated. Flower heads develop into dense, bristly seed heads containing small, round, black seeds enclosed in a brown sheath.

GERMINATION AND REGENERATION: Seeds germinate readily on bare sunny soil in late spring or early summer. In poorly maintained agricultural fields, hundreds of seedlings per square yard (meter) can germinate and provide serious competition for crops.

HABITAT PREFERENCES: As befits a species lacking mycorrhizae, redroot pigweed grows best in rich soil and full sun. Like many warm-season grasses, the amaranths utilize C4 photosynthesis, which enables them to flourish under hot, dry conditions.

CULTURAL SIGNIFICANCE: Seeds and leaves of various amaranth species were an important source of food and medicine for Native Americans. Some people still gather the young leaves and shoots for spring greens. Prior to the Spanish conquest of South and Central America, an amaranth species known as *kiwicha* (*Amaranthus caudatus*) was an important cultivated grain for the Aztecs and the Incas. In more recent times, Roundup-resistant varieties of a closely related species, *Amaranthus palmeri,* have begun appearing in fields where the herbicide has been heavily used in the cultivation of genetically modified soybeans.

RELATED SPECIES: **Prostrate pigweed (***Amaranthus blitoides* **S. Wats.)** is a low-growing, C4 annual that is native to south central North America. It forms a spreading mat with succulent, smooth, green or red stems up to a foot (30 cm) or so tall; its hairless leaves are up to 1.5 inches (3.7 cm) long with a prominent notch at their tip. Prostrate pigweed produces small, greenish flowers from June through August in dense clusters at the ends of the branches and in the lower leaf axils. It is common in a variety of disturbed sunny sites, and its seeds were once used as a food source by various Native American tribes, including the Zuni.

Taproot of redroot pigweed

Redroot pigweed growth habit

Redroot pigweed (*left*) and prostrate pigweed (*right*)

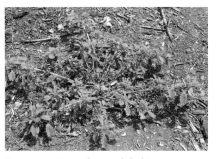

Redroot pigweed flowers

Prostrate pigweed produces leaves with a notched tip

Prostrate pigweed growth habit

Chenopodium album L.   Common Lambsquarters

Synonyms: fat hen, mealweed, goosefoot, bacon-weed, wild spinach

Life Form: summer annual; up to 6 feet (1.8 m) tall

Place of Origin: Eurasia and North Africa

Vegetative Characteristics: Stems have conspicuous grooves and are usually green, but sometimes show some purple coloration at the point where the branches attach. The plant branches freely and develops a broad, pyramidal shape at maturity. Alternate leaves are dull green, 2–4 inches (5–10 cm) long, roughly triangular to rhomboidal (hence the common name goosefoot) with irregular teeth, and gray to white, "mealy" undersides. Lambsquarters has a fleshy consistency and sometimes turns purplish at the end of the growing season; it produces a short, tenacious taproot.

Flowers and Fruit: Large, pale green inflorescences terminate the branches in late summer; wind-pollinated flowers lack petals and are inconspicuous. Fruits are tiny, bladderlike structures that contain a single seed. A large plant can produce up to 75,000 seeds.

Germination and Regeneration: Seeds fall to the ground at maturity and germinate in late spring to early summer; they are also eaten and dispersed by ground-feeding birds. Buried seeds can remain viable in the soil for decades, if not centuries.

Habitat Preferences: Lambsquarters tolerates a wide variety of soil, moisture, and light regimens but reaches its full potential in nutrient-rich ground. The plant is noteworthy for its ability to remain green after other plants have "browned out" from drought or frost. It is common in all sorts of disturbed sites, including neglected ornamental landscapes, minimally maintained public parks, vacant lots, rubble dumps, small pavement openings, chain-link fence lines, stone walls, median strips, and railroad rights-of-way. It does not form mycorrhizae.

Ecological Functions: Food and habitat for wildlife; phytoremediation of heavy metals (zinc, copper, and lead) on contaminated land.

Cultural Significance: Young lambsquarters shoots are edible in the spring after the fine powder that typically covers the leaves is washed away. In times of famine in Europe, the seeds were boiled to make gruel or baked into bread; it is also cultivated as a food plant in northern India. The recently popular *quinoa* is the seed of *Chenopodium quinoa*, domesticated by the Incas and cultivated by their descendants at high elevations in the Andes.

Related Species: Mexican tea or **wormseed (*Dysphania ambrosioides* (L.) Mosyakin & Clemants)** is an upright or sprawling plant, native to South, Central, and North America. It grows best in full sun and can reach heights up to 3 feet (0.9 m) tall; it produces bright green, ovate to lanceolate leaves with wavy margins that are 1–3 inches (2.5–7.5 cm) long by 0.5–1.5 inches (1.2–3 cm) wide. The foliage emits a pungent camphor-like odor when crushed, and the seeds have a long history of use to expel worms, especially in children. Mexicans call the plant *epazote* and add it to chili sauces and bean dishes to aid digestion and reduce flatulence.

Lambsquarters in full flower

Lambsquarters flowers

Lambsquarters growing between a sidewalk and a storm drain

Lambsquarters foliage

Mexican tea in flower

Mexican tea growth habit

Daucus carota L.   Wild Carrot

SYNONYMS: Queen Anne's lace, bird's nest

LIFE FORM: **herbaceous biennial**; up to 3–4 feet (1–1.3 m) tall

PLACE OF ORIGIN: Eurasia and North Africa

VEGETATIVE CHARACTERISTICS: Wild carrot is a tall, slender plant with finely dissected, pinnately compound foliage that has an aromatic, carrot-like odor. During its first year, the plant forms a rosette of bipinnately compound leaves—up to 6 inches (15 cm) long—that remain green through the winter; the second year it sends up a tall flowering stalk with alternate leaves. The stout, whitish taproot is difficult to pull out of the ground.

FLOWERS AND FRUIT: Wild carrot produces numerous lace-like white flowers in flat-topped, terminal clusters (*umbels*) from June through September; they can be either insect- or self-pollinated. Approximately one in four plants has a single, deep purple flower (the "fairy seat") in the center of the cluster of all-white flowers. As the seeds develop, the umbels close up and develop a form resembling a bird's nest. The tiny seeds are covered with numerous barbs that facilitate their dispersal by animals. A single plant can produce up to 4,000 seeds, and the tall stalks are often bent over by their weight.

GERMINATION AND REGENERATION: Seeds germinate readily on disturbed, sunny sites in spring.

HABITAT PREFERENCES: Wild carrot tolerates full sun and dry soil. It is common in abandoned grasslands and urban meadows, vacant lots, rubble dumps, rock outcrops, stone walls, roadsides, and railroad rights-of-way. In its native European habitat it is common in coastal meadows.

ECOLOGICAL FUNCTIONS: Disturbance-adapted colonizer of bare ground; tolerant of roadway salt and compacted soil; food for wildlife.

CULTURAL SIGNIFICANCE: Seeds of wild carrot have long been used in European traditional medicine as a "morning-after" contraceptive, and in India to reduce female fertility. Indeed, Dioscorides' first-century herbal, *De Materia Medica*, clearly describes its anti-fertility properties. The use of wild carrot as a contraceptive has been documented in the Appalachian Mountains of North Carolina as well, passed down through oral tradition (Riddle 1999). The plant has also been used as a diuretic to cure kidney and bladder stones and to eliminate worms. Although it is considered the ancestor of the domestic carrot, the roots are barely edible. The plant has never been popular with farmers: Connecticut required farmers to control the plant in 1822, and Darlington and Thurber (1859, 147) interpreted the presence of this plant in one's fields as a sign of moral weakness: "When it gets on the premises of a careless, slovenly farmer, it soon multiplies so as to become a source of annoyance to the whole neighborhood."

Wild carrot growth habit

Wild carrot flowering on a roadside

Wild carrot foliage

Wild carrot flower head with "fairy seat" in the center

Developing seeds of wild carrot

Apocynum cannabinum L.   Dogbane Hemp

SYNONYMS: Indian hemp, hemp dogbane, Indian physic

LIFE FORM: **herbaceous perennial**; up to 5 feet (1.5 m) tall

PLACE OF ORIGIN: North America

VEGETATIVE CHARACTERISTICS: Dogbane hemp produces smooth, reddish stems that are branched in their upper half, and all parts of the plant exude a milky sap when broken. Its opposite leaves have distinct petioles and are variable in shape—from lanceolate to ovate—and are 2–3 inches (5–7.5 cm) long by 0.5 to 1.0 inch (1.2–2.5 cm) wide. In moist, rich soil the plant can be quite tall and erect; in poor, dry soil it is shorter and more spreading.

FLOWERS AND FRUIT: Small, white or greenish-white flowers are produced in erect, terminal clusters in June and July. Individual flowers possess both male and female parts, are less than 0.25 inch (6 mm) across, and are pollinated by insects. It produces pairs of long, thin fruits, 4–6 inches (10–15 cm) long, which change color from green to red as they age. At maturity, the pods split open to release numerous small seeds topped with a tuft of silky hairs that facilitate their dispersal by wind.

GERMINATION AND REGENERATION: Seeds germinate in sunny, open areas in the spring; established plants form extensive clones made up of shoots produced by its widely spreading root system.

HABITAT PREFERENCES: Dogbane hemp grows in full sun in well-drained, gravelly or sandy soil on the edges of woods and fields as well as in marshy areas near the banks of streams and rivers; it is also tolerant of dry soil and partial shade. In urban areas, it can be found in disturbed waste areas.

ECOLOGICAL FUNCTIONS: Tolerant of compacted soil; stream and riverbank stabilization; food and habitat for wildlife.

CULTURAL SIGNIFICANCE: Native Americans harvested the roots of dogbane hemp at the end of the growing season for a variety of medicinal purposes; they also used the stems—after the plant had died back—as a source of strong fibers for making rope and other utilitarian products, hence its common name. All parts of the plant, including the milky sap it exudes when damaged, are toxic.

RELATED SPECIES: **Spreading dogbane (*Apocynum androsaemifolium* L.)** resembles dogbane but is not as erect in its growth habit, typically growing 2–4 feet (0.6–1.2 m) tall. Its bell-shaped flowers are more than twice as wide and showier than those of dogbane hemp, with pink, recurved petals conspicuously marked with reddish stripes. Like dogbane hemp, its roots have a history of medicinal use by Native Americans and its milky sap is toxic. It is more of an upland species than is dogbane hemp, and it is tolerant of poor, sandy soils.

Dense stand of dogbane hemp in Detroit

Flowers and foliage of dogbane hemp

Close-up of dogbane hemp flowers

Fruits and foliage of dogbane hemp

Growth habit and foliage of spreading dogbane

Flowers of spreading dogbane

Asclepias syriaca L.   Common Milkweed

Synonyms: *Asclepias cornuti*, wild cotton, silkweed, cotton weed

Life Form: **herbaceous perennial**; up to 3–4 feet (0.9–1.2 m) tall

Place of Origin: eastern North America

Vegetative Characteristics: Common milkweed produces stout, erect stems that are 3–4 feet (0.9–1.2 m) tall, hollow, and unbranched. Leaves are opposite, oblong to oval, and 3–6 inches (7.5–15 cm) long with a prominent white midvein; the upper surface of the leaf is smooth, the underside is covered with downy hair. All parts of the plant are covered with soft hairs and exude milky sap when broken.

Flowers and Fruit: Milkweed produces its fragrant, 5-petaled, pink-purple to white flowers from late June through early August. The distinctive round, drooping inflorescence attracts a variety of insect pollinators, including monarch butterflies, who drink the nectar they produce but do not actually transfer any pollen. Fruits are 3–4 inch (7.5–10 cm) long, teardrop-shaped pods with prominently curved tips, soft spines, and a gravity-defying upward orientation. In October they split open to release hundreds of flattened, black seeds, each topped with a tuft of soft, silky fibers (*floss*) that facilitate wind dispersal.

Germination and Regeneration: Seeds germinate readily in open, sunny locations. Established plants produce new shoots from the base of the old stalks and also sprout from buds produced by the widely spreading roots, forming extensive colonies over time.

Habitat Preferences: Milkweed grows best in sunny, disturbed habitats without much regard for soil pH. It is common in infrequently mowed meadows, sandy roadsides, vacant lots, and along railroad tracks.

Ecological Functions: Milkweed produces cardenolide compounds that are toxic to many animals because of their impact on heart function. Insects that feed on the leaves of the plant, including caterpillars of the monarch butterfly, sequester these cardenolides in their bodies, making themselves and the butterflies they become toxic to any bird that might try to eat them, thereby discouraging predation. Numerous other plants in the Apocynaceae family also produce toxic cardenolides, including dogbane hemp (*Apocyanum*), oleander (*Nerium*), periwinkles (*Vinca* and *Catharanthus*), and swallowwort (*Vincetoxicum*).

Cultural Significance: Young milkweed shoots—less than 8 inches (20 cm) tall—are edible when cooked. Native Americans made a tea from the roots as a laxative and for treating kidney stones and dropsy; made rope from its stem fibers; and made pillows from its floss. During World War II, the silky seed hairs were used as a substitute for South American kapok to fill "Mae West" life vests. Between 1943 and 1945, a million such flotation devices were filled with the floss from some 24 million pounds (11 million kg) of milkweed pods! Milkweed was introduced into Europe at an early date, and Carl Linnaeus, father of the binomial nomenclature system, mistakenly thought the plant originated in Syria, hence the species name *syriaca*.

Milkweed can spread
extensively from root
sprouts

Milkweed flowers

Milkweed foliage

Milkweed seed dispersal

Milkweed fruits

SYNONYMS: *Cynanchum nigrum, C. lousisae,* climbing milkweed, Louise's swallow-wort, black dog-strangling vine

LIFE FORM: **herbaceous perennial vine**; 3–6 feet (1–2 m) long

PLACE OF ORIGIN: Europe

VEGETATIVE CHARACTERISTICS: The smooth, unbranched stems of black swallowwort can either twine up through and around other vegetation or sprawl along the ground. Simple, opposite leaves are 2–4 inches (5–10 cm) long, glossy, dark green, and lance-shaped with short petioles. Unlike other members of the milkweed family, the stems and leaves do not exude milky latex when broken.

FLOWERS AND FRUIT: Black swallowwort produces clusters of small, purple-black flowers with 5 triangular petals from June through September. The flowers, which can be either self- or fly-pollinated, are followed by pairs of slender, green seedpods that are 2–3 inches (5–7.5 cm) long and taper to a sharp point. Seedpods contain numerous seeds tufted with silky white hairs that facilitate wind dispersal.

GERMINATION AND REGENERATION: Seeds germinate in semi-shade or full sun and develop into perennial crowns that generate numerous shoots early in the spring. Pulling out these shoots without removing the buried crown has little effect on the plant's ability to survive and spread.

HABITAT PREFERENCES: This drought-adapted species can tolerate a wide variety of environmental conditions from full sun to dense shade. It grows in disturbed or neglected urban habitats, including residential and commercial landscapes, minimally maintained public parks and open spaces, vacant lots, urban meadows, rock outcrops, stone walls, roadsides, and on chain-link fences where it is particularly common. Plants often get established at the base of ornamental shrubs and can be virtually impossible to remove.

CULTURAL SIGNIFICANCE: Black swallowwort was first reported in North America in the mid-1800s in Massachusetts, presumably an escape from the Harvard Botanical Garden. Its distribution in the Northeast has increased dramatically since the 1970s, and many states now list it as an invasive species. The plant is toxic and has been used in European traditional medicine as a laxative, diuretic, emetic, and anti-tumor agent.

RELATED SPECIES: **Pale swallowwort** or **dog-strangling vine** (*Vincetoxicum rossicum* **(Kleopov) Barbar.**) is similar to black swallowwort in all respects except that its flowers are pale yellow to pink-purple. It is less widespread than black swallowwort on the East Coast but is more common in the Great Lakes Basin. It was reported growing in Toronto, Canada, in 1889. The Greek physician Dioscorides included the plant in his first-century herbal compendium, *De Materia Medica.*

Black swallowwort seedpods about to open

Black swallowwort flowers

Pale swallowwort flowers

Black swallowwort's underground root and shoot system

Black swallowwort on a chain-link fence

Black swallowwort dispersing its seeds

Black swallowwort along a curb

Achillea millifolium L.   Yarrow

SYNONYMS: milfoil, noble yarrow, staunchweed, soldier's woundwort, green arrow, nosebleed, sanguinary, yarroway, herbe militaris, knight's milfoil

LIFE FORM: herbaceous perennial; up to 3 feet (0.9 m) tall

PLACE OF ORIGIN: Europe, Asia, and North America

VEGETATIVE CHARACTERISTICS: Yarrow is an upright plant with hairy stems and alternate, pinnately compound leaves up to 6 inches (15 cm) long; leaves are subdivided multiple times, giving them a finely dissected appearance. A basal rosette of leaves typically stays green through the winter; the fibrous root system is extensive but shallow.

FLOWERS AND FRUIT: Yarrow produces flat-topped, terminal clusters of white or pink flowers throughout the summer. Individual flower heads are less than 0.25 inch (6 mm) in diameter and consist of 5 "petals" (ray florets) surrounding 10 or more yellow disk florets that are pollinated by insects. At maturity the seed heads contain hundreds of tiny, brown seeds that are dispersed as the plant blows back and forth in the wind. The spent flower stalks persist through the winter, making the dormant plant easy to identify.

GERMINATION AND REGENERATION: Seeds germinate readily on bare ground; established plants spread by underground rhizomes.

HABITAT PREFERENCES: Yarrow is an extremely drought-tolerant species that grows well in sunny, dry soil. It is common in minimally maintained public parks, vacant lots, rubble dumps, urban meadows, unmowed highway banks, and roadsides. In its native habitat, yarrow grows in grasslands and open woodlands.

ECOLOGICAL FUNCTIONS: Tolerant of roadway salt and compacted soils; erosion control on slopes; food and habitat for wildlife.

CULTURAL SIGNIFICANCE: Tea made from flowering yarrow plants has a long history of use in European and Chinese folk medicine for treating colds, fevers, and inflammation; the crushed leaves have been used in poultices for dressing wounds to promote blood coagulation. According to Greek mythology, Achilles (for whom the plant is named) learned the medicinal value of yarrow from the centaur Chiron and used the plant to heal his soldiers' wounds while fighting in Troy. The plant is listed in Dioscorides' first-century herbal, *De Materia Medica*. Native Americans used yarrow, which is also native to North America, to treat numerous ailments and as a sacred sweat-lodge herb. Yarrow is widely cultivated as a drought-tolerant perennial, and horticultural selections are available in a variety of colors including red, pink, yellow, gold, and white. In the past, landscape architecture students used the tops of dead flower stalks as trees in their models.

Stand of yarrow

Yarrow foliage

Yarrow growth habit

Yarrow in full flower

Close-up of
yarrow flower
heads

Ageratina altissima (L.) King & H.E. Robins.   White Snakeroot

Synonyms: *Eupatorium rugosum, Ageratum altissimum,* white sanicle, richweed

Life Form: herbaceous perennial; 2–4 feet (60–120 cm) tall

Place of Origin: eastern and central North America

Vegetative Characteristics: White snakeroot produces oppositely arranged, dark green leaves that are broadly ovate to lanceolate in shape with 3 main veins. They can reach up to 6 inches (15 cm) long by 4 inches (10 cm) wide with petioles up to 2 inches (5 cm) long; they are hairless, coarsely toothed, and taper to a sharp point. Leaves become smaller, less coarsely toothed, and more lanceolate the higher up they are on the stem.

Flowers and Fruit: White snakeroot produces flat-topped *panicles* of snowy white flowers—2–6 inches (5–12 cm) across—from late summer through frost. The inflorescences terminate the branches and consist of numerous flower heads that contain 10–25 star-shaped disk florets, each approximately 0.12 inch (3 mm) across; there are no ray florets. Following pollination by insects, small wind-dispersed seeds develop. Because of its late bloom period, the plant only becomes conspicuous in September.

Germination and Regeneration: Seeds germinate readily under a variety of conditions, and established plants can form large colonies from rhizomes. Under ideal conditions, the plant can spread quickly and extensively.

Habitat Preferences: White snakeroot grows best in full to partial shade in reasonably good, moist to dry woodland soil. It is most common along the edges of disturbed woods or open thickets but can also be found in urban habitats, such as shady vacant lots, backyards, and the edges of walkways.

Cultural Significance: The foliage and rhizomes of white snakeroot contain a fat-soluble alcohol known as *tremetol.* Americans who settled in the Ohio River Valley in the early 19th century and drank milk from cows that had grazed on white snakeroot developed an often fatal disease called milk sickness or "trembles." Abraham Lincoln's mother, Nancy Hank Lincoln, supposedly died of milk sickness in 1818. The plant is more common west of Pennsylvania than it is to the east. The plant's common name is derived from a report that Native Americans used a decoction of the roots as a remedy for snakebite.

Related Species: White Snakeroot can be confused with **white wood aster** (*Eurybia divaricata*), which blooms at the same time and in the same type of habitat, but it has daisy-like flowers with ray florets. White snakeroot might also be confused with two native white-flowered thoroughworts that bloom in the fall: **tall thoroughwort** (*Eupatorium altissimum* L.) and **late thoroughwort** (*E. serotinum* **Michx.**). White snakeroot produces leaves with broader blades and longer petioles than either of these species, and it prefers woodland shade whereas the thoroughworts grow in sunnier, more open habitats.

White snakeroot growing in a shady backyard

White snakeroot leaves

Branching structure of white snakeroot inflorescences

White snakeroot growth habit

White snakeroot flower heads and star-shaped florets

Dense stand of white snakeroot under the shade of black walnut trees

Ambrosia artemisiifolia L.   Common Ragweed

Synonyms: *Ambrosia elatior, A. media,* hayfever weed, wild tansy, bitter weed, hogweed, wild wormwood

Life Form: summer annual; up to 4 feet (1.3 m) tall

Place of Origin: North America

Vegetative Characteristics: Stems are upright, hairy, and rough to the touch; the highly dissected, compound leaves are 2–4 inches (5–10 cm) long and give ragweed a lacy appearance. Leaves on the lower portion of the stem are opposite; those on the upper part are mostly alternate. The plant remains green until late in the growing season.

Flowers and Fruit: Yellow-green male flowers are arranged in conspicuous terminal spikes that occupy the top one-third of the plant and resemble miniature candelabras. They produce copious amounts of highly allergenic, wind-dispersed pollen in September and October. The inconspicuous female flowers are located in leaf axils lower on the stem and produce 1-seeded fruits.

Germination and Regeneration: Seeds germinate readily during late spring on bare ground once the temperature is consistently above 50° F (10° C). Seeds can remain viable in the soil for up to 80 years.

Habitat Preferences: Ragweed grows equally well in dry, sandy soil or heavy, moist soils with a neutral or higher pH. It is common in disturbed habitats including roadsides with heavy salt applications (which elevates soil pH); vacant lots and rubble dumps; small-scale pavement openings and cracks; chain-link fence lines; rock outcrops and crumbling stone walls; and compacted lawns where mowing reduces it to a stub that still manages to flower despite being only 2 inches (5 cm) tall.

Ecological Functions: Disturbance-adapted colonizer of bare ground; food and habitat for wildlife (especially birds); phytoremediation of soils contaminated with lead.

Cultural Significance: Ragweed is notorious for producing abundant pollen that causes hay fever in many people in late summer and early fall. Plants grown in a controlled environment with elevated CO_2 produced roughly a third more pollen than did plants grown in ambient air, suggesting that ragweed may become a more serious allergy problem in the future (Ziska et al. 2003). In 1759, Philadelphia naturalist John Bartram described the plant's affinity for disturbance: "Ye lesser ambrosia is a very troublesome weed in plantations where it hath got ahead. It is an annual & grows with corn & after harvest it shoots above ye stuble growing 3 or 4 foot high & so thick that one can hardly walk through. It tastes very bitter & if milch cows feeds upon it (for want of enough grass) their milk will taste very loathsome. It seldom grows to any head next year nor until ye field is plowed or sowed again" (p. 453).

Ragweed foliage

Growth habit of flowering ragweed

Ragweed dominating the disturbed edge of a Connecticut bike path

Repeatedly mowed ragweed flowering at 3 to 4 inches (7.5–10 cm) tall

Male flowers of ragweed

Spiny clusters of female ragweed flowers located below male flowers

Arctium minus (Hill) Bernh.   Common Burdock

SYNONYMS: *Lappa minor*, clotbur, cuckoo-button, lesser burdock, Velcro plant

LIFE FORM: **biennial**; up to 6 feet (2 m) tall

PLACE OF ORIGIN: Eurasia

VEGETATIVE CHARACTERISTICS: Burdock produces a rosette of large evergreen leaves in its first year and a tall, erect, multibranched flower stalk in the second year. Alternate, heart-shaped leaves can be up to 20 inches (50 cm) long by 16 inches (40 cm) wide; their upper surface is dark green and smooth, the undersurface is light green and woolly. The stout, fleshy taproot makes the plant difficult to remove.

FLOWERS AND FRUIT: Terminal clusters of small, red-purple, thistle-like flower heads—consisting entirely of disk florets—are produced at the tips of all the branches at the end of burdock's second year. Flower heads, which are produced from July through October, are 0.75–1.5 inches (1.8–3.6 cm) wide and, following pollination by insects, develop into a bur covered with spiny, hooked bracts that attach and hold tenaciously to animal fur and human clothing.

GERMINATION AND REGENERATION: Burdock seeds germinate readily in disturbed sunny or shady locations with moist soil; burdock has no vegetative regeneration.

HABITAT PREFERENCES: Burdock is found in a wide variety of habitats but grows best in moist, rich soil in full or half sun. It is common in minimally maintained public parks; vacant lots and rubble dump sites; the edges of emergent woodlands; the sunny borders of freshwater wetlands, ponds, and streams; and on unmowed highway banks and median strips with frequent salt applications.

CULTURAL SIGNIFICANCE: Burdock has a long history of use in traditional medicine in both Europe and China: a tea from the root is used as a "blood purifier" and to treat gout, rheumatism, and liver and kidney ailments; seeds are used to treat various skin eruptions. The Greek physician Dioscorides included the plant in his first-century herbal, *De Materia Medica*. The root from first-year plants are edible when boiled; indeed, the Japanese vegetable *gobo* comes from domesticated varieties of burdock's close relative, *Arctium lappa*. Flower stalks, when peeled and chopped, are edible after boiling. Burdock was an early arrival in North America and is listed in Josselyn's *New-England's Rarities Discovered,* published in 1672. In the Shakespeare play *As You Like It*, burdock is a prominent metaphor in a conversation between Rosalind and Celia, who notes with sadness, "O how full of briars is this work-a-day world." To which the cheerful Rosalind responds, "They are but burs, cousin, thrown upon thee in holiday foolery. If we walk not in the trodden paths, our very petticoats will catch them." The melancholy Celia replies, "I could shake them off my coat; these burs are in my heart." And finally, the Swiss engineer George de Mestral came up with the idea for Velcro while studying burdock burs under the microscope while removing them from his dog after a hike in the Alps.

Burdock foliage at the end of its first growing season

Burdock bolting into flower

Mature burdock fruits at the end of its second season

Burdock flower heads

Developing burdock seed heads can be used as a brooch

153

Artemesia vulgaris L.   *Mugwort*

SYNONYMS: chrysanthemum weed, common or wild wormwood

LIFE FORM: herbaceous perennial; up to 6 feet (2 m) tall

PLACE OF ORIGIN: Eurasia

VEGETATIVE CHARACTERISTICS: Mugwort is a tall, upright plant that develops thick, woody stems by late in the growing season. Leaves are alternate, 2–4 inches (5–10 cm) long, and deeply dissected with sharp-pointed lobes. The upper surface of the leaf is smooth; the underside is covered with woolly white hairs, giving the plant a gray-green appearance, especially after it begins to flower in August. The foliage gives off a pungent odor when crushed because of the chemical *thujone*. Mugwort is a taxonomically complex species that shows considerable variation, probably as a result of hybridization with closely related species.

FLOWERS AND FRUIT: Mugwort produces a feathery terminal inflorescence in late summer and fall made up of numerous small, greenish-yellow, composite flower heads that consist entirely of wind-pollinated disk florets. The weight of the developing fruits typically causes the plant to arch over by the end of the season.

GERMINATION AND REGENERATION: Mugwort seeds are quite small and seem to be dispersed mainly by wind but also by water. Seedlings are extremely competitive and readily colonize existing meadows; once established, plants spread vigorously by rhizomes and typically take over the area. Buried seeds remain viable in the soil for many years.

HABITAT PREFERENCES: Mugwort is the quintessential urban weed. It thrives on disturbed, compacted soil with high pH and is common in minimally maintained public parks, vacant lots, rubble dumps, soil stockpiles, small pavement openings and cracks, and along railroad tracks. It recently has begun to spread along highway margins and median strips in suburban and rural areas, most likely in response to heavy applications of road salt.

ECOLOGICAL FUNCTIONS: Phytoremediation in degraded urban landscapes (absorbs heavy metals zinc, copper, lead, and cadmium and binds them to organic matter); soil building on degraded soil.

CULTURAL SIGNIFICANCE: Mugwort was used to flavor beer before hops took over that role (hence its common name). In traditional Chinese medicine, the compressed, dried leaves are burned on the skin to stimulate acupuncture points (moxa) and to treat rheumatism. In European traditional medicine, a tea made from mugwort leaves was used to treat epilepsy; menstrual, menopausal, and gastrointestinal problems; to increase the flow of urine; and to stimulate the appetite. European and Asian cuisines use the young leaves to flavor a variety of traditional dishes. The plant was introduced into North America at an early date, undoubtedly for medicinal purposes (Barney 2006). Mugwort pollen is a potent allergen and can be a cause of fall hay fever. The Ukrainian city of Chernobyl, made uninhabitable by the explosion of its nuclear power plant in 1986, was named after the mugwort that once grew in the area and is still common.

Stand of mugwort in May (note silvery leaf undersides)

Mugwort leaf shape is highly variable

Mugwort dominating a neglected urban sidewalk

Mugwort foliage in early spring

Mugwort in flower in September

155

Bidens frondosa L.   Devil's Beggarticks

SYNONYMS: sticktights, devil's bootjack, cuckold, beggarticks, bur-marigold

LIFE FORM: **summer annual**; up to 3 feet (0.9 m) tall

PLACE OF ORIGIN: eastern North America

VEGETATIVE CHARACTERISTICS: Devil's beggarticks is an upright, loosely branched plant with 4-angled, often purplish stems. Opposite, compound leaves have 3–5 large, lance-shaped leaflets up to a foot (30 cm) long with serrated edges. Leaves at the top of the stem and close to the flower heads are mostly simple. In fall, leaves often turn purple.

FLOWERS AND FRUIT: Devil's beggarticks produces green and yellow flower heads— about 1 inch (2.5 cm) wide—at the ends of its branches from August through October. Flower heads consist of numerous disk florets with golden-yellow corollas (there are no ray florets) surrounded by 8 or so narrow, leaf-like outer bracts that are approximately 1.5 inches (3.7 cm) long. Although visited by insects, the flowers are mainly self-pollinated and give rise to flattened, brown fruits (*achenes*) about 0.5 inch (1.2 cm) long with a pair of slender, barb-covered spikes (awns) at their apex. The whole structure looks something like an old-fashioned boot jack.

GERMINATION AND REGENERATION: Barbed seeds (actually fruits) stick tenaciously to human clothing and to animal fur making them both effective dispersal agents; in addition, seeds can be blown around by the wind. They germinate in spring under a wide variety of conditions.

HABITAT PREFERENCES: Devil's beggarticks grows in habitats ranging from dry and shady to moist and sunny but does best in moist soil and full sun. It is common in minimally maintained public parks; woodlands that develop on abandoned open space; the edges of freshwater wetlands, ponds, and streams; drainage ditches; and landscape planting beds.

CULTURAL SIGNIFICANCE: Native Americans made a tea from the leaves to expel worms and chewed the leaves for sore throat. The Shakers sold the plant as an expectorant and to treat heart palpitations and "uterine derangement," and to induce sweating, menstruation, and urination.

Devil's beggarticks growth habit in flower

Devil's beggarticks foliage

Devil's beggarticks typically grows in moist, shady conditions

Devil's beggarticks flower heads

Devil's beggarticks developing seed head

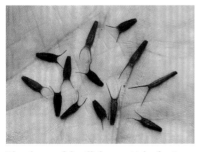

The shape of devil's beggarticks fruits are the source of the common name, devil's bootjack

Centaurea stoebe Tausch   Spotted Knapweed

Synonyms: *Centaurea maculosa, C. biebersteinii*, star thistle

Life Form: biennial or **herbaceous perennial**; 1–4 feet (30–120 cm) tall

Place of Origin: Europe

Vegetative Characteristics: Spotted knapweed forms a basal rosette of deeply lobed, gray-green leaves approximately 6 inches (15 cm) long during its first year. In its second year, it produces slender, ribbed stems with alternate, gray-green leaves that are pinnately dissected with numerous lobes; leaves near the top of the plant are tiny and narrow.

Flowers and Fruit: Spotted knapweed produces thistle-like flower heads that are about 0.5–1 inch (1.2–2.5 cm) wide at the ends of all the branches from mid-to-late summer. Each flower head consists of numerous pink to purple ray florets surrounded by a base of overlapping, grayish-green floral bracts (*phyllaries*) that taper to a black tip with short bristles. A variety of insects, including honeybees and butterflies, pollinate the flowers. The dry fruits (*achenes*) have short bristles that facilitate animal dispersal, but they are also dispersed by wind.

Germination and Regeneration: Seeds germinate readily in fall or spring in dry, sandy soil in full sun; mature plants produce short rhizomes that give rise to new shoots.

Habitat Preferences: Spotted knapweed grows well in disturbed sites with dry, low-fertility soils. It is common in vacant lots and along roadways and railroad tracks. The plant is highly salt tolerant and common near the coast. Spotted knapweed produces allelopathic chemicals that suppress the growth of nearby plants of a different species, one of the reasons for its success as an invasive species.

Ecological Functions: Disturbance-adapted colonizer of bare ground; food and habitat for wildlife.

Cultural Significance: Spotted knapweed is a fairly recent, unintentional introduction into North America, having first been recorded in the 1890s. It has since become a major problem in the rangelands of Idaho, Montana, and eastern Oregon and Washington. Federal and state agencies have spent millions of dollars trying to control spotted knapweed, and many states list it as an invasive species.

Spotted knapweed in a typical urban habitat

Spotted knapweed growth habit

Spotted knapweed foliage in spring

Spotted knapweed in flower

Spotted knapweed flower head

Cichorium intybus L.   Chicory

SYNONYMS: succory, blue sailors, ragged sailors, blue daisy, coffee-weed, wild endive

LIFE FORM: herbaceous perennial; up to 5 feet (1.6 m) tall

PLACE OF ORIGIN: Eurasia

VEGETATIVE CHARACTERISTICS: Chicory forms a basal rosette of dark green, lance-shaped, alternate leaves with coarsely toothed margins that resemble those of dandelion. Basal leaves are 3–9 inches (7.5–22.5 cm) long by 1–3 inches (2.5–7.5 cm) wide; upper leaves are much smaller and their bases clasp the stem. Chicory forms a deep, perennial taproot, and all parts of the plant exude milky sap when broken.

FLOWERS AND FRUIT: Chicory's tall flower stalks emerge from basal rosettes from July through October; the bright blue (and occasionally pink) flower heads, which are approximately 1.5 inches (3.7 cm) across, open in the morning and close by afternoon. The individual florets that make up the flower head are all of the ray type with a single, attached, blue "petal," and they are commonly pollinated by bees.

GERMINATION AND REGENERATION: Birds eat and disperse the dried fruit, and the seeds germinate in a variety of sunny sites, especially those with high soil pH; established plants sprout back from the base in early spring.

HABITAT PREFERENCES: Chicory is a common roadside plant that grows well in sunny, dry habitats and in soils with an elevated pH. In the urban environment, it is common in minimally maintained public parks, the margins of neglected residential and commercial landscapes, vacant lots, rubble dumps, pavement cracks, chain-link fence lines, unmowed highway banks, median strips, and railroad rights-of-way.

ECOLOGICAL FUNCTIONS: Tolerant of roadway salt and compacted soil; food and habitat for wildlife; erosion control on slopes; soil building on degraded land.

CULTURAL SIGNIFICANCE: Domesticated varieties of chicory, known as Belgium endive, have distinctive spoon-shaped, pale white leaves that are produced by growing year-old plants in the dark. This technique, known as blanching, prevents the new leaves from turning green and makes them less bitter than they would be if they were grown in the sun. Radicchio is a red-leaved variety of chicory that has been cultivated in the Mediterranean region for thousands of years. The root of chicory, when roasted and ground, has long been used as a coffee additive or substitute, especially during times of scarcity. The root has also been used in traditional European medicine to treat liver problems, gout, sour stomach, and rheumatism. Chicory has more recently been evaluated as a forage species for feeding livestock; for phytomining and phytoremediation purposes (based on its capacity to absorb heavy metals); and as a commercial source of *inulin*, which is used as a low-calorie sweetener by the food industry. The leafy salad greens escarole and curly endive (*Cichorium endivia* L.) are often confused with chicory.

Chicory in full flower

Chicory foliage rosettes before flowering

Chicory growth habit as its flower stalks are expanding

Chicory flourishing along a crumbling cement walkway

Chicory flowers

Cirsium vulgare (Savi) Tenore   Bull Thistle

SYNONYMS: *Carduus vulgaris, Cirsium lanceolatum*, spear thistle

LIFE FORM: biennial; up to 6 feet (1.8 m) tall

PLACE OF ORIGIN: Eurasia and North Africa

VEGETATIVE CHARACTERISTICS: During its first year, bull thistle forms a rosette of spiny, dark green leaves approximately 8 inches (20 cm) long that persists through the winter. The following spring the rosette "bolts" and produces a tall, upright stalk. Leaves are alternate, with sharp spines on their lobes and bases that fuse with the stem, giving it a winged or fluted appearance.

FLOWERS AND FRUIT: Bull thistle produces reddish-purple flower heads at the ends of the stalks from June through October. These heads, which are approximately 1.5 inches (3.7 cm) wide, consist entirely of disk florets; their bases are covered with sharp-tipped floral bracts. Following insect- or self-pollination, a spiny capsule develops that is filled with small seeds, each topped with a downy pappus.

GERMINATION AND REGENERATION: Wind-dispersed seeds germinate readily in full sun in a variety of habitats and soil types. Plants show no form of vegetative regeneration.

HABITAT PREFERENCES: Bull thistle grows best in moist, rich or clayey soil in full sun. In the urban environment, it is common in disturbed habitats including rubble-covered vacant lots, grasslands with compacted soil, the rocky edges of freshwater wetlands and streams, unmowed highway banks and median strips, and in the ballast along railroad tracks.

CULTURAL SIGNIFICANCE: When fully mature, this biennial is a dramatic presence in the landscape. Young, elongating stems are edible when peeled and boiled. The plant was inadvertently introduced into North America sometime in the 1700s as a contaminant in seed or hay. Thistles have always had image problems, dating back to the Old Testament days when God banished Adam from the Garden of Eden with the words, "cursed is the ground because of you. . . . It will produce thorns and thistles for you" (Genesis 3, verses 17–18).

RELATED SPECIES: Canada thistle (*Cirsium arvense* (L.) Scop.) is a perennial species from Eurasia that grows to approximately 4 feet (1.2 m) tall. It produces numerous pink-purple flower heads about an inch (2.5 cm) across throughout the summer; individual flowers are insect-pollinated (honey bees), and wind-dispersed seeds mature in August. It grows best in disturbed, sunny sites in rich or clay soils and produces vigorous new shoots from deep, creeping roots, eventually forming large colonies. Canada thistle is difficult to eradicate and was one of the first plants that state governments (Vermont in 1791, New York in 1812) required farmers to control (Pauly 2007). The plant has a long history of medicinal use in Europe as a tonic, a diuretic, and an astringent for treating skin sores and rashes.

Bull thistle flower and seed heads

Bull thistle foliage

Bull thistle leaf rosette after its first year of growth

Canada thistle growth habit

Bull thistle flower head

Canada thistle flower heads

Erechtites hieracifolius (L.) Raf. ex DC.   Fireweed

SYNONYMS: pilewort, American burnweed, white fireweed

LIFE FORM: **summer annual**; up to 6 feet (2 m) tall

PLACE OF ORIGIN: eastern North America

VEGETATIVE CHARACTERISTICS: Fireweed has smooth, unbranched, bright green stems and leaves. Leaves are alternate, oblong to lance-shaped with sharp, irregular teeth along their margins; as the plant matures the leaves become increasingly lobed—almost pinnate—and can reach lengths up to 8 inches (20 cm) by 3 inches (7.5 cm) wide.

FLOWERS AND FRUIT: Fireweed flower heads have a tight, tubular form, and the individual disk florets barely extend above the upright green bracts. Dozens of these composite flowers are typically arranged into a terminal, flat-topped cluster. Flower heads are self- or insect-pollinated and quickly mature into "puffball" seed heads composed of small black fruits topped with a cluster of slender, soft white hairs (*pappus*) that facilitate wind dispersal.

GERMINATION AND REGENERATION: These wind-dispersed seeds are famous for their ability to germinate on land that has been recently burned, but fireweed also grows on sites that have been disturbed by other forces (see below). Less well appreciated is that seeds can germinate after years of burial, when some form of physical disturbance brings them to the surface.

HABITAT PREFERENCES: In the urban environment, fireweed grows in a variety of disturbed sites in full sun, typically as a single plant or in small groups but occasionally in large patches. It is common in small pavement openings, along the margins of walls and chain-link fences, in minimally maintained landscape plantings, and along railroad tracks. Its appearance in the landscape—often together with pokeweed—is generally an indication of soil disturbance.

CULTURAL SIGNIFICANCE: Native Americans and early European settlers used fireweed to treat hemorrhoids (piles), and the Shakers sold an extract of the plant to treat "diseases of the mucous tissues of the lungs, stomach and bowels." John Bartram published a remarkably detailed account of the ecology of fireweed in 1759: "We have another weed called Cotton groundsel which grows with us 6 or 7 foot high & ye stalk at bottom near as thick as my wrist. In our new cleared land after ye first plowing in ye spring or in our marshes ye year after they are drained it grows there all over so close that there is no passing along without breaking it down to walk or ride through it but in ould fields or medows there is not one stalk to be seen now. If we put ye question how comes this to grow so prodigiously on ye new land plowed ground & perhaps not one root growing within several miles ye answer is very ready. . . . One day when ye sun shined bright a little after its meridian my Billy [Bartram's son] was looking up at it when he discovered an innumerable quantity of downey motes floating in ye air between him & ye sun. He immediately called me out of my study to see what thay were. . . . Some lowered & fell into my garden where we observed every particular detachment of down spread in 4 or 5 rays with a seed of ye grounsel in its center. How far these was carried by that breeze cant be known but I think thay must have come near five miles from a meadow to reach my garden" (p. 453).

Fireweed growth habit in fall

Foliage of a fireweed seedling

Fireweed flower heads

Fireweed seeds ready for dispersal

Fireweed dispersing its seeds

Dissected leaves of fireweed

Erigeron annuus (L.) Pers.   Annual Fleabane

SYNONYMS: daisy fleabane, white top

LIFE FORM: **annual** or **biennial**; up to 3–4 feet (0.9–1.2 m) tall

PLACE OF ORIGIN: eastern North America

VEGETATIVE CHARACTERISTICS: Annual fleabane is an upright plant with ridged, hairy stems and alternate leaves with toothed margins. Leaves at the base of the stem are up to 4 inches (10 cm) long, elliptical or egg-shaped with distinct petioles; those at the top are linear or lance-shaped with short or no petioles.

FLOWERS AND FRUIT: Annual fleabane produces terminal clusters of daisy-like flower heads, approximately 0.5–0.75 inch (1.2–1.8 cm) in diameter, made up of 50–120 thin, white "petals" (ray florets) surrounding a yellow center (disk florets). The plant begins blooming in early summer and, depending on the weather, into the fall. Flowers are commonly visited by insects, but recent research indicates that the small, wind-dispersed seeds are produced asexually and are genetically identical with the plant that produced them—a form of clonal reproduction known as *apomixis*.

GERMINATION AND REGENERATION: Seeds germinate during spring in sunny, disturbed sites.

HABITAT PREFERENCES: Annual fleabane grows best in fertile soil in full sun but also tolerates dry soil. In the urban environment, it is common along the margins of minimally maintained public parks; in vacant lots, rubble dumps, abandoned lawns, urban meadows, and unmowed highway banks and median strips; and along railroad tracks.

CULTURAL SIGNIFICANCE: Native Americans used annual fleabane to treat a variety of ailments. Burning it is said to drive away fleas and gnats (hence the common name), but no evidence supports this claim. The species has become weedy in Europe, reversing the typical Europe to North America scenario.

Annual fleabane growth habit

Field of annual fleabane

Annual fleabane in full bloom

Annual fleabane growing between sidewalk and street

Annual fleabane flower head

Erigeron canadensis L.   Horseweed

Synonyms: *Conyza canadensis*, Canada fleabane, mare's tail, butter weed

Life Form: winter or **summer annual**; up to 6 feet (1.8 m) tall

Place of Origin: North America

Vegetative Characteristics: Horseweed produces a single, tall, hairy stem that branches only at the top when it flowers in July and August. Alternate, sessile leaves are approximately 4 inches (10 cm) long by 0.5 inch (1.2 cm) wide, slightly toothed; they are crowded along the stems such that they appear almost whorled. Both the stem and the leaves are covered with long, spreading, white hairs. Seedlings form a basal rosette of leaves that deteriorates as the stem elongates in early summer. Before horseweed flowers, it can easily be mistaken for a goldenrod except that it produces only a single stem.

Flowers and fruit: Horseweed produces masses of tiny, white flowers at the ends of all the upper branches from July through October. Individual flower heads consist of numerous small "petals" (ray florets) surrounding the yellow disk florets, which can be self- or insect-pollinated. The weight of the developing seeds gives the plant its top-heavy appearance and often causes it to bend over or break.

Germination and regeneration: The tiny horseweed seeds are topped with whitish bristles (*pappus*) that facilitate wind dispersal; they can germinate on a wide variety of disturbed sites. Seeds that germinate in late summer form rosettes that overwinter and bolt in the summer; those that germinate in spring bolt in the fall.

Habitat Preferences: Horseweed grows well in dry, exposed sites in full sun. It prefers rich, loamy soil, but it also does well on soils with high clay or gravel content. Horseweed is common in minimally maintained public parks and residential landscapes, vacant lots, rubble dumps, abandoned grasslands, rock outcrops, stone walls, fence lines, small pavement openings, railroad tracks, and unmowed highway banks and median strips.

Ecological Functions: Disturbance-adapted colonizer of bare ground; food and habitat for wildlife.

Cultural Significance: Native Americans used horseweed to treat diarrhea, and the young leaves are edible after boiling. The Shakers sold it to treat kidney stones and a variety of bowel and urination problems. Horseweed has become a major invasive species in parts of temperate Asia and Europe.

Stand of flowering horseweed in Boston

Horseweed in early
stages of flowering

Horseweed growing along railroad tracks

Close-up of horseweed flower heads

Horseweed prior to
flowering

Galinsoga quadriradiata Ruiz & Pav.   Hairy Galinsoga

SYNONYMS: *Galinsoga ciliata*, quick weed, gallant soldier, galinsoga

LIFE FORM: **summer annual**; up to 2 feet (60 cm) tall

PLACE OF ORIGIN: Central America

VEGETATIVE CHARACTERISTICS: Hairy galinsoga produces dark green, densely branched stems that are completely covered with coarse hairs; leaves are opposite, oval to triangular, coarsely toothed, and 1–3 inches (2.5–7.5 inches) long. The root system is fibrous, shallow, and relatively easy to pull up.

FLOWERS AND FRUIT: The tiny flower heads of hairy galinsoga, which are about 0.25 inch (6 mm) wide, are produced from June through frost. They consist of 5 white "petals" (*ray florets*) with notched tips surrounding a central core of yellow disk florets. They can be either insect- or self-pollinated, and they produce tiny seeds (*achenes*) with a tuft of papery scales that facilitates their dispersal.

GERMINATION AND REGENERATION: Galinsoga seeds germinate throughout the summer on disturbed soil. A single plant can produce multiple generations in a season, making for explosive growth in midsummer. Buried seeds retain their viability in the soil for many years.

HABITAT PREFERENCES: In full sun and fertile soil, hairy galinsoga can become quite large and robust. In the urban environment, it is found in neglected residential and commercial landscapes, minimally maintained public parks and open spaces, vacant lots, rubble dumps, small pavement openings, and sidewalk cracks. The plant is noteworthy for its ability to complete its life cycle from germination to seed production in as little as 3–4 weeks.

ECOLOGICAL FUNCTIONS: Disturbance-adapted colonizer of bare ground; food and habitat for wildlife.

CULTURAL SIGNIFICANCE: Hairy galinsoga seems to have arrived in the Northeast in the 1920s, having migrated up from Central America and through the Southeast; it is now ubiquitous throughout eastern North America. It was introduced into Europe from Peru in the late 1700s and is now considered an invasive species across most of the continent; it has also spread to Asia. In spring, the leaves and stems of hairy galinsoga are edible when cooked.

RELATED SPECIES: **Smooth galinsoga (*Galinsoga parviflora* Cav.)** is native to South and Central America and is similar to hairy galinsoga except that its stems are smooth or only slightly hairy.

Hairy galinsoga growth habit

Hairy galinsoga in its typical urban habitat

Hairy galinsoga foliage and flowers

Hairy galinsoga flower heads

Close-up of hairy galinsoga flower heads

Hieracium sabaudum L.   New England Hawkweed

SYNONYMS: Savoy hawkweed. *Note:* Even experts have trouble distinguishing this highly variable European species from another European species, ***Hieracium lachenalii,*** and the native North American ***Hieracium canadense.***

LIFE FORM: **herbaceous perennial**; up to 3 feet (0.9 m) tall

PLACE OF ORIGIN: Eurasia

VEGETATIVE FEATURES: New England hawkweed produces erect stems with alternate, dark green, lance-shaped leaves that are 3–5 inches (7.5–12.5 cm) long. Leaves have sharp-pointed teeth scattered along the margins and are hairy on the underside. All parts of the plant exude milky sap when broken. Unlike many other hawkweeds, this upright species does not form a basal rosette.

FLOWERS AND FRUIT: Conspicuous, bright yellow flower heads, similar to those of dandelion, are produced from August through October. The flower heads, which are produced at the ends of long, highly branched stems, are approximately an inch (2.5 cm) wide and consist entirely of ray florets. They can be self- or insect-pollinated and develop into round, "puffball" fruiting heads about 0.75 inch (1.8 cm) wide that are composed of numerous, small, wind-dispersed seeds.

GERMINATION AND REGENERATION: Seeds germinate best in moist sites in partial sun. Established plants produce basal buds that can give rise to dense clusters of new stems. Unlike many other hawkweeds, this species is not rhizomatous.

HABITAT PREFERENCES: New England hawkweed grows best in dry, shady sites with compacted or sandy soil, but it can also tolerate full sun. In the Northeast, it is much more common in the urban environment than in the countryside. It is found especially along the shady margins of disturbed or emergent woodlands, in the rain shadow of buildings, and around the base of rock outcrops.

CULTURAL SIGNIFICANCE: New England hawkweed appears to have been introduced into the Boston area in the late 1800s and now appears to be spreading throughout the Northeast.

RELATED SPECIES: **Field hawkweed** (***Hieracium caespitosum* Dumort.** [formerly *H. pratense*]) is another perennial European species that spreads by rhizomes. It produces rosettes of lance-shaped leaves that are conspicuously hairy on both sides and 2–10 inches (5–25 cm) long. In June and July, it produces a leafless flower stalk that is 1–3 feet (30–90 cm) tall and topped with 4 or more yellow, dandelion-like flowers about an inch (2.5 cm) wide. Field hawkweed grows in sunny, nutrient-poor, sandy soils. **Mouse-eared hawkweed** (***Pilosella officinarum* Vaill.** [formerly *Hieracium pilosella*]) is a prostrate herbaceous perennial from Europe that produces a basal rosette of oblong leaves 4 inches (10 cm) long by 0.5 inch (1.2 cm) wide. Leaves are green above and whitish below and covered with long, stringy hairs. In summer, it produces solitary, golden-yellow flower heads at the top of a hairy, leafless stalk up to 14 inches (35 cm) tall. It grows in full sun and poor soil (including lawns) and reproduces by runners that give rise to large colonies that suppress the growth of nearby plants.

New England hawkweed growth habit in early summer

New England hawkweed inflorescence

New England hawkweed in flower at the edge of a woodland path

New England hawkweed flower and seed heads

Mouse-eared hawkweed in flower

Field hawkweed in flower

Lactuca serriola L.   Prickly Lettuce

SYNONYMS: *Lactuca scariola*, compass plant

LIFE FORM: **summer annual** or **biennial**; up to 6 feet (1.8 m) tall

PLACE OF ORIGIN: Europe

VEGETATIVE CHARACTERISTICS: Prickly lettuce produces a basal rosette of pale green leaves that typically gives rise to a single hollow, upright, green to white stem. Leaves are alternate, 2–10 inches (5–25 cm) long, and usually (but not always) deeply lobed with rounded, half-moon–shaped sinuses; the upper surface of the leaf is smooth, but a single row of stiff prickles is conspicuous along the midrib of the lower surface. The name "compass plant" can be attributed to leaves that are typically turned on edge and oriented vertically toward the sun. All parts of the plant exude milky sap when broken (hence the Latin name *Lactuca*).

FLOWERS AND FRUIT: Yellow flower heads are less than 0.25 inch (6 mm) wide and are grouped into a large, pyramidal panicle that terminates the main stem. Flower heads consist entirely of ray florets and are produced from July through September; they can be either self- or insect-pollinated. Seeds develop rapidly, and each has a feathery pappus to aid in its dispersal by wind.

GERMINATION AND REGENERATION: Seeds germinate readily in a variety of soils and exposures.

HABITAT PREFERENCES: Prickly lettuce grows best in nutrient-rich soils and full sun but can tolerate dry sites with poor soil. In the urban environment, it is common in neglected residential and commercial landscapes, minimally maintained public parks, unmowed highway banks and median strips, and small pavement openings and cracks.

ECOLOGICAL FUNCTIONS: Prickly lettuce is a phytoremediation plant because of its ability to absorb heavy metals (mainly zinc and cadmium) into its tissues.

CULTURAL SIGNIFICANCE: The wild ancestor of cultivated lettuce, this plant has been used in traditional European medicine as a diuretic and to stimulate the flow of milk in nursing mothers. This latter use is based on the ancient *Doctrine of Signatures* theory, which holds that herbs resembling a particular body part are useful for treating ailments of that body part. Young shoots can be eaten but are not recommended given that the bitter, milky sap they contain has mild narcotic properties and, when dried, was once used as a substitute for opium.

Prickly
lettuce
with entire
(unlobed)
leaves

Prickly lettuce with lobed leaves (note
orientation perpendicular to the sun)

Prickly lettuce
in its typical
urban habitat
between a
sidewalk and
a wall

Note the stiff prickles along the underside of
prickly lettuce leaf

Prickly lettuce flower heads

Prickly lettuce seed heads

Leucanthemum vulgare Lam.   Oxeye Daisy

SYNONYMS: *Chrysanthemum leucanthemum*, white daisy, white weed, marguerite, snow-in-June

LIFE FORM: herbaceous perennial; up to 2 feet (70 cm) tall

PLACE OF ORIGIN: Europe

VEGETATIVE CHARACTERISTICS: In early spring, oxeye daisy produces a basal rosette of hairless, dark green, alternate leaves—up to 6 inches (15 cm) long—with rounded lobes and long petioles. As the flower stalks elongate in late spring, the plant produces sessile leaves that become progressively smaller, narrower, and less lobed toward the top.

FLOWERS AND FRUIT: This species produces classic daisy "flowers" at the ends of long stalks from late May through July. Insect-pollinated flower heads are 1–2 inches (2.5–5 cm) across with 20–30 white "petals" (ray florets) surrounding a cluster of yellow disk florets. Fruits (achenes) are tiny and lack the bristles (*pappus*) that facilitate wind dispersal in many other members of the Aster Family.

GERMINATION AND REGENERATION: Small, dark-brown seeds are dispersed by wind and birds. Buried seeds retain their viability in the soil for many years. Established plants can form large clumps that increase in size through the growth of rhizomes.

HABITAT PREFERENCES: Oxeye daisy is highly drought tolerant and grows best in soils with an elevated pH and full sun. It is common in compacted lawns, minimally maintained public parks, vacant lots, rubble dumps, urban meadows, rock outcrops, stone walls, unmowed highway banks, median strips, and railroad rights-of-way.

ECOLOGICAL FUNCTIONS: Tolerant of roadway salt and compacted soil; food for wildlife; soil building on degraded land; erosion control on slopes.

CULTURAL SIGNIFICANCE: Most people assume this common, attractive species is native to North America, but it was introduced from Europe in the early 1700s, probably as a contaminant in hay brought over to feed livestock. The plant has a long tradition of medicinal use in Europe to treat whooping cough, asthma, and nervous excitability. Oxeye daisy is a component in many of the wildflower seed mixes used to stabilize disturbed landscape situations. The young leaves and unopened flower buds can be eaten raw in salads. Philadelphia naturalist John Bartram was inspired to become a botanist when he paused in the middle of plowing his fields to study an oxeye daisy: "What a shame, said my mind, or something that inspired my mind, that thee shouldst have employed so many years in tilling the earth and destroying so many flowers and plants, without being acquainted with their structures and their uses!" (Crèvecœur 1783, 181).

Stand of
oxeye daisy
in a degraded
urban meadow

Oxeye daisy foliage rosette in early
spring

Oxeye daisy growth habit
just before flowering

Oxeye daisy flower heads

Close-up of oxeye daisy flower head

Senecio vulgaris L.   Common Groundsel

SYNONYMS: ground glutton, chickenweed, grimsel, ragwort

LIFE FORM: **winter** or **summer annual**; up to 18 inches (45 cm) tall

PLACE OF ORIGIN: Eurasia

VEGETATIVE CHARACTERISTICS: Groundsel produces smooth, green stems that are initially erect but, as they branch, they bend over and form large, spreading clumps. Matte-green leaves are alternate, smooth, have irregularly toothed or deeply lobed margins, and are 6–10 inches (15–25 cm) long with a semi-succulent texture.

FLOWERS AND FRUIT: The terminal clusters of groundsel flower heads are approximately 0.5 inch (1.2 cm) wide and consist entirely of yellow disk florets (no ray florets); they are mainly self-pollinated and quickly mature to produce small, brown seeds topped with a fluffy white pappus that aids wind dispersal. The plant is typically in flower and fruit simultaneously.

GERMINATION AND REGENERATION: Seeds germinate in both spring and fall, when the weather is cool. Seeds can survive burial in the soil for many years.

HABITAT PREFERENCES: Groundsel grows best in moist, nutrient-rich soils in spring and fall. It is common in a wide variety of urban habitats, including small pavement openings, stone walls, rock outcrops, vacant lots, waste dumps, moist lawn areas, and ornamental landscape plantings.

CULTURAL SIGNIFICANCE: Groundsel, collected in spring, has a long history of medicinal use in Europe as a diuretic and a purgative. The plant was an early arrival in North America and is listed in Josselyn's *New-England's Rarities Discovered* (1672) under the category: "Of such plants as have sprung up since the English planted and kept cattle in New England." It contains alkaloids that can poison livestock.

RELATED SPECIES: **Pineapple weed (*Matricaria discoidea* DC.)** is a winter or summer annual up to 18 inches (45 cm) tall that grows best in full sun and is an indicator of sandy or compacted soils; its stems are smooth and green and can be either erect or prostrate depending on growing conditions. It is extremely tolerant of trampling. Shiny, bright green leaves are finely dissected, alternate, and 0.5–2 inches (1.2–5 cm) long. All of its parts have a distinct pineapple-like scent when crushed. Greenish-yellow flower heads—0.25–0.5 inch (6–12 mm) wide—look like cone-shaped buttons and lack ray florets; they are produced from May through September followed by tiny, brown fruits that lack a feathery pappus. Flowers are visited by insects but are mostly self-pollinated. Pineapple weed seems to be native to both northeastern Asia and northwestern North America. Young flower heads, either dried or fresh, make a refreshing tea and a tea made from the leaves has been used in traditional medicine for stomachaches and colds.

Groundsel flower and seed heads

Groundsel growth habit

Close-up of groundsel in flower
and seed heads

Pineapple weed growth habit

Pineapple weed approximately
4 inches (10 cm) tall

Pineapple weed flower heads

Solidago canadensis L.   Canada Goldenrod

SYNONYMS: yellow-weed, common goldenrod

LIFE FORM: **herbaceous perennial**; up to 5 feet (1.6 m) tall

PLACE OF ORIGIN: eastern North America

VEGETATIVE CHARACTERISTICS: Canada goldenrod is a tall, erect plant with smooth or slightly hairy stems and alternate, linear to lance-shaped leaves that are tapered at both ends. Most leaves are sharply serrated along their margins and can reach up to 6 inches (15 cm) long by 0.75 inch (1.8 cm) wide.

FLOWERS AND FRUIT: Canada goldenrod produces conspicuous yellow flowers arranged in curved, pyramidal clusters from August through early October. Individual flower heads consist of small, yellow ray florets surrounding equally small, yellow disk florets, which produce sticky pollen that attracts a wide variety of insects, including honeybees, bumblebees, beetles, and butterflies. The tiny seeds have a few short bristles at one end (*pappus*) that facilitate wind dispersal.

GERMINATION AND REGENERATION: Seeds germinate in sunny sites with moist or dry soil. Creeping rhizomes grow out from the main stem in late summer, which give rise to a cluster of genetically identical plants in spring. The plant spreads aggressively and can form large, persistent colonies over time.

HABITAT PREFERENCES: Canada goldenrod grows best in well-drained soil and full sun. In the urban environment, it is commonly found in neglected residential and commercial landscapes, minimally maintained public parks, vacant lots, rubble dump sites, and abandoned grasslands and meadows. In rural areas, it can take over abandoned fields and pastures, unmowed highway banks, drainage ditches, and railroad rights-of-way.

ECOLOGICAL FUNCTIONS: Soil-building colonizer of bare ground; food and habitat for wildlife (especially pollinators); erosion control on slopes.

CULTURAL SIGNIFICANCE: The conspicuous, insect-pollinated goldenrods are often blamed for the allergies caused by the much less conspicuous, wind-pollinated ragweed, which blooms at the same time. Native Americans used all parts of goldenrod for a variety of medicinal purposes. Thomas Edison experimented with goldenrod sap to produce rubber but was unable to develop a commercial product from it. Canada goldenrod has become a major invasive species in temperate Asia and Europe.

RELATED SPECIES: Canada goldenrod is highly variable and is difficult to distinguish from two closely related species that can grow to 7 feet (2.1 m) tall: **late goldenrod** (*S. gigantea* Ait.) has lance-shaped, sharply toothed leaves and smooth, green or purplish stems often covered with a whitish bloom; and **tall goldenrod** (*S. altissima* L.) has leaves that are rough-textured on their upper surface and a downy, grayish stem.

Canada goldenrod growth habit

Canada goldenrod rhizomes

Canada goldenrod in full flower

Canada goldenrod in a meadow on
Spectacle Island in Boston Harbor

Mature seeds of Canada goldenrod

Solidago sempervirens L.   Seaside Goldenrod

Synonym: salt-marsh goldenrod

Life Form: herbaceous perennial; 4–6 feet (1.2–1.8 m) tall

Place of Origin: East Coast of North America from Newfoundland to the Gulf of Mexico

Vegetative Characteristics: Seaside goldenrod is an erect or sprawling plant with smooth stems and smooth, toothless, semi-succulent dark green leaves. They are mostly sessile and arranged alternately along the stem; those making up the basal rosette are large—up to 10 inches (25 cm) long by 1–2 inches (2.5–5 cm) wide—and often persist over the winter. No other *Solidago* in our region produces fleshy leaves like those of seaside goldenrod.

Flowers and Fruit: Seaside goldenrod is easy to recognize in the landscape because it blooms later than other goldenrods in our area, from mid-September well into October. The multibranched, terminal inflorescence is golden-yellow, relatively dense, and quite showy; individual flowers are significantly larger than those of other goldenrods, about 0.25 inch (6 mm) wide with numerous ray flowers less than 0.12 inch (3 mm) long. Because of its late bloom time, seaside goldenrod is an important food source for many pollinators, most notably fall-migrating monarch butterflies. Small seeds mature in November and are topped with short bristles (*pappus*) that facilitate wind dispersal.

Germination and Regeneration: Seeds germinate in a variety of sunny, well-drained sandy or gravelly soils; established plants sprout from the base to form dense clumps.

Habitat Preferences: Seaside goldenrod is native to Atlantic coastal areas where its tolerance of saline soils and salt spray make it common on salt marshes and sand dunes. Away from the coast, it is locally common along the margins of highways and in the infrastructure cracks on bridges and concrete embankments.

Cultural Significance: Although native to coastal areas, seaside goldenrod has been spreading rapidly inland since the 1980s following heavy applications of road salt along major highway corridors (Brauer and Gerber 2002). It is an example of a native species whose range has been expanding in response to changing environmental conditions.

Related Species: Early goldenrod (*Solidago juncea* Aiton) is easy to identify in the landscape because it blooms well before the other goldenrods, beginning in July and continuing through August. Its smooth, green stems grow to approximately 4 feet (1.2 m) tall with a terminal, plume-like panicle of yellow, insect-pollinated flowers that branch outward like a fireworks' display. Early in the growing season, the plant produces smooth, lanceolate leaves at its base that can be large—up to 8 inches (20 cm) long by 1.5 inches (3.7 cm) wide—with a distinct petiole, whereas leaves higher up on the stem are much smaller. Early goldenrod spreads by rhizomes and can form large colonies; it grows best in full sun and is tolerant of dry soil.

Seaside goldenrod in its native habitat on Spectacle Island in Boston Harbor

Seaside goldenrod inflorescences

Seaside goldenrod growth habit

Seaside goldenrod growing along a heavily salted highway in Boston

Early goldenrod inflorescence

Early goldenrod in bloom

Sonchus oleraceus L.   Annual Sowthistle

SYNONYMS: common sowthistle, milk thistle, hare's lettuce, hare's thistle

LIFE FORM: summer annual; up to 4 feet (1.2 m) tall

PLACE OF ORIGIN: Europe

VEGETATIVE CHARACTERISTICS: Annual sowthistle is an unbranched plant with smooth, erect stems and foliage that has a waxy bluish appearance. Deeply lobed leaves are alternately arranged, up to a foot (30 cm) long, and have margins that are irregularly toothed and slightly prickly; leaf lobes near the base of the petiole seem to clasp or wrap around the stem. All parts of the plant exude milky white sap when broken.

FLOWERS AND FRUIT: Terminal clusters of pale yellow flower heads 0.5–1 inch (1.2–2.5 cm) in diameter are produced from July through October. Flower heads consist entirely of ray florets and, following self- or insect-pollination, develop into tiny, white "puffballs" consisting of brown seeds with a feathery pappus that facilitates wind dispersal.

GERMINATION AND REGENERATION: Seeds germinate in a wide variety of conditions from sun to shade and moist to dry.

HABITAT PREFERENCES: Annual sowthistle grows best in rich soil in full sun. It is common in neglected residential and commercial landscapes, minimally maintained public parks, small pavement openings and cracks, vacant lots, and rubble dumps.

CULTURAL SIGNIFICANCE: Dioscorides included annual sowthistle in his first-century herbal, *De Materia Medica*. The species name, *oleraceus*, means "of the vegetable garden" in Latin, and indeed, the plant is an edible potherb in early spring. Rabbits seem to be particularly fond of common sowthistle, which is also known as hare's lettuce. The plant was an early arrival in North America and appears in Josselyn's *New-England's Rarities Discovered* (1672) under the category: "Of such plants as have sprung up since the English planted and kept cattle in New England."

RELATED SPECIES: Perennial sowthistle (*Sonchus arvensis* L.) grows to approximately 4 feet (1.3 m) tall and sends up new shoots from a persistent rhizome. Its leaves resemble those of annual sowthistle with toothed edges and rounded lobes; those at the base of the plant are about 12 inches (30 cm) long by 2 inches (5 cm) wide, and they shrink in size as they ascend the stem. In July and August, it produces bright yellow, dandelion-like flower heads that are quite showy and nearly twice as big as those of annual sowthistle—up to 2 inches (5 cm) wide. This species is much more common in the countryside than in the city. Young shoots are edible in spring.

Annual sowthistle growth habit

Annual sowthistle leaf

Annual sowthistle flower heads

Mature seed head of annual sowthistle

Close-up of
perennial
sowthistle
flower head

Perennial sowthistle in flower

SYNONYMS: *Aster pilosus*, frost aster, awl aster

LIFE FORM: **herbaceous perennial**; up to 3 feet (0.9 m) tall

PLACE OF ORIGIN: eastern and central North America

VEGETATIVE CHARACTERISTICS: White heath aster's numerous erect stems—which can be either hairy or smooth—sprout from a perennial crown to form a hemispherical clump. The lower leaves, which are about 4 inches (10 cm) long and lance-shaped with smooth margins, are typically shed before flowering begins in the fall. Smaller, heath-like leaves on the upper portions of the stem are less than 0.5 inch (1.2 cm) long and persist through flowering.

FLOWERS AND FRUIT: From late August or early September through October, white heath aster is covered with masses of small, white (occasionally pink), daisy-like flower heads in a showy, dome-shaped display. They are about 0.5 inch (1.2 cm) across and are composed of numerous "petals" (ray florets) with yellow centers (disk florets) that turn reddish following pollination by a variety of insects. The tiny seeds have numerous bristles at the apex that facilitate wind dispersal.

GERMINATION AND REGENERATION: Seeds germinate on bare soil in sunny locations; established plants sprout from a perennial woody crown.

HABITAT PREFERENCES: White heath aster grows best in full sun in dry, sandy soils but can also tolerate compacted soils. It is common in disturbed sites including unmanaged urban meadows, vacant lots, rubble dump sites, small pavement openings and cracks, and unmowed highway banks.

CULTURAL SIGNIFICANCE: White heath aster is a remarkably tough native species that is strikingly beautiful in flower. It tolerates urban conditions and high soil pH and blooms very late in the season—features that make it a good candidate for cultivation in low-maintenance meadows. *Aster* means "star" in Latin.

RELATED SPECIES: Heart-leaved aster (*Symphyotrichum* (*Aster*) *cordifolium* (L.) Nesom) is a perennial native species that produces upright clusters of pale blue-violet flower heads—0.5–0.75 inch (1.2–1.8 cm) across—at the ends of the branches from August through October. Yellow disk flowers become reddish following pollination. The plant grows to 3 feet (0.9 m) tall and has heart-shaped leaves with conspicuously toothed margins. Heart-leaved aster is remarkably drought tolerant and is common in semi-shaded urban habitats at the edges of roadways, woodlands, vacant lots, and neglected gardens. **White wood aster** (*Eurybia* (*Aster*) *divaricata* (L.) Nesom) is a native species that produces flat or rounded clusters of white flower heads—0.75–1 inch (1.8–2.5 cm) across—from August through September. It grows to approximately 3 feet (0.9 m) tall and has heart-shaped, coarsely toothed leaves. Common in semi-shade at the edges of country roads and in disturbed urban woodlands, white wood aster spreads vigorously from rhizomes and is conspicuous when in bloom.

White heath aster in an urban meadow in September

White wood aster in the shade of an oak in September

White heath aster in full bloom

Heart-leaved aster foliage

White heath aster flower heads

Heart-leaved aster in full bloom in October

Tanacetum vulgare L.   Common Tansy

SYNONYMS: *Chrysanthemum uliginosum*, golden buttons, buttons

LIFE FORM: **herbaceous perennial**; up to 4 feet (1.2 m) tall

PLACE OF ORIGIN: Europe

VEGETATIVE CHARACTERISTICS: Tansy is an upright plant with smooth, 5-angled stems. Alternate, pinnately compound leaves are 4–8 inches (10–20 cm) long and half as wide; margins of the leaflets are deeply toothed, producing a fern-like appearance. Bruised foliage is highly aromatic and smells something like camphor.

FLOWERS AND FRUIT: Tansy produces golden-yellow inflorescences in a terminal, flat-topped cluster from July through September. Individual flower heads are button-shaped, about 0.5 inch (1.2 cm) wide, and consist entirely of insect-pollinated disk florets. The small seeds that follow lack any obvious dispersal mechanism, such as a pappus or barbs.

GERMINATION AND REGENERATION: Tansy seeds germinate in a variety of habitats; established plants produce short rhizomes that develop into extensive, long-lived clumps.

HABITAT PREFERENCES: Tansy is a grassland or meadow species in its native Europe, where it is tolerant of compacted soil, high pH, and full sun. In the urban environment, it is common in minimally maintained public parks and open spaces, vacant lots, rubble dumps, urban meadows, and unmowed highway banks.

ECOLOGICAL FUNCTIONS: Disturbance-adapted colonizer of bare ground; soil building on degraded land; erosion control on slopes; food and habitat for wildlife.

CULTURAL SIGNIFICANCE: Tea or oil made from tansy leaves has long been used in Europe to treat dyspepsia, flatulence, jaundice, sore throat, and as a wash for swellings and inflammations. For culinary purposes, the leaves and shoots are added to a variety of puddings and omelets, and are used to make "tansy cakes," which are traditionally eaten at Easter. Tansy was cultivated at John Winthrop's Plymouth colony in New England as early as 1631, and Darlington and Thurber (1859, 190) noted that it "was originally introduced as a garden-plant, and generally cultivated for its aromatic bitter properties—which have rendered it a prominent article in the popular *Materia Medica*. It has now escaped from the gardens, and is becoming naturalized—and something of a weed—in many places." In the past, landscape architecture students collect the dried flower heads to use as trees in their models.

Tansy growth habit

Tansy foliage

Close-up of tansy flowering heads

Tansy flowers attract pollinators

Spent seed heads of tansy

Taraxacum officinale Weber ex Wiggers   Dandelion

SYNONYMS: *Leontodon taraxacum*, lions-tooth, blow-ball, cankerwort, pissabed

LIFE FORM: herbaceous perennial; leaves up to 8 inches (20 cm) long

PLACE OF ORIGIN: Eurasia

VEGETATIVE CHARACTERISTICS: Dandelion is a low-growing plant that produces a basal rosette of long, dark leaves that are 3–10 inches (7.5–25 cm) long and have deeply lobed margins whose tips point back toward the crown of the plant. The stout, fleshy taproot can be difficult to remove. All parts of the plant exude milky sap when broken.

FLOWERS AND FRUIT: Dandelion produces bright yellow flower heads in early spring and, to a lesser extent, in fall. The heads are produced at the ends of smooth, hollow stalks and consist entirely of petal-like ray florets that are 1–2 inches (2.5–5 cm) wide. Although often visited by insects seeking pollen and nectar, seeds usually form *apomictically* (without the benefit of sexual union) before turning into distinctive globe-shaped "puffballs."

GERMINATION AND REGENERATION: Wind-dispersed seeds germinate readily under a wide variety of ecological conditions. Because of the absence of functional sexual reproduction, seedlings are genetically identical to the parent that produced them. Dandelion sprouts readily from a perennial crown as well as from pieces of the root left behind following failed attempts at removal.

HABITAT PREFERENCES: Dandelion tolerates a wide range of growing conditions. It is common in lawns, neglected ornamental landscapes, vacant lots, rubble dumps, small pavement openings, rock outcrops, stone walls, highway banks and median strips, drainage ditches, and railroad rights-of-way. In Europe, dandelion is found in sunny, disturbed habitats with limestone soils.

CULTURAL SIGNIFICANCE: Dandelion was an early arrival in North America as documented by Josselyn in *New-England's Rarities Discovered* (1672). Young leaves are widely collected for salad greens in early spring (they are rich in vitamins and minerals), dandelion wine can be made from the fermented flowers, and in fall the roots can be roasted and used as a coffee substitute. Tea made from the fresh root acts as a diuretic and has long been used in Europe to treat liver, bladder, and kidney ailments. American homeowners trying to grow the perfect lawn hate dandelions and spend millions of dollars on herbicides to control it. This behavior is quite a turnaround when one considers an early guidebook to New York's Central Park that described "blessed dandelions in such beautiful profusion as we have never seen elsewhere, making the lawns in places, like green lakes reflecting a heaven sown with stars" (Pauly 2007, 177).

RELATED SPECIES: Coltsfoot (*Tussilago farfara* L.) produces yellow, dandelion-like flower heads in early spring, before its leaves emerge. The upper surface of its distinctive, hoof-shaped leaves is smooth and dull green; the lower is covered with woolly, white hairs that are soft to the touch. It is more of a country than a city plant but can often be found in the ballast gravel along railroad tracks. It has a long history of medicinal use in Europe to treat asthma and to suppress coughing.

Dandelion in early spring

Dandelion seed head

Dandelions are among the first plants to bloom in early spring

Dandelion foliage rosette in fall

Dandelion being visited by flies

Coltsfoot foliage and mature seed heads

Tragopogon pratensis L.　Meadow Salsify

SYNONYMS: Meadow goat's beard, showy goat's beard, wild oyster plant, jack-go-to-bed-at-noon

LIFE FORM: biennial; up to 3 feet (0.9 m) tall

PLACE OF ORIGIN: Eurasia

VEGETATIVE CHARACTERISTICS: Meadow salsify forms a basal rosette of linear, grass-like leaves in its first year. In its second season, the plant produces hollow, flowering stems that are smooth, round, and semi-succulent. Dull green leaves are approximately a foot (30 cm) long and uniformly narrow with bases that clasp and enclose the stem. The fleshy taproot and all other parts of the plant exude a milky sap when broken.

FLOWERS AND FRUIT: Meadow salsify produces solitary yellow flower heads—up to 2.5 inches (6.2 cm) wide—at the ends of tall flowering stalks in late spring and early summer. Flower heads consist entirely of petal-like ray florets subtended by numerous sharp-pointed, green bracts that are equal in length to the petals of the ray florets. Flower heads turn conspicuously toward the sun and typically remain open until midday (hence one of its common names). Following pollination by insects, bracts continue to elongate, eventually forming an elongated cone that encloses the developing seeds. Mature seed heads look like those of dandelion but are much larger—up to 3 inches (7.5 cm) across—and brown rather than white.

GERMINATION AND REGENERATION: Wind-dispersed seeds can travel more than 800 feet (250 m) from their parent and germinate in sunny, disturbed sites.

HABITAT PREFERENCES: Meadow salsify grows best in full sun and dry soil. It is common in abandoned grasslands, urban meadows, vacant lots, rubble dumps, and at the base of rock outcrops and stone walls.

CULTURAL SIGNIFICANCE: The white root of meadow salsify, if collected before the flowering stalk develops, is edible when cooked. Young stems and bases of the lower leaves can also be eaten after boiling. Garden salsify or oyster plant (*Tragopogon porrifolius*) has purplish flowers and produces a large, white root that can be eaten throughout the winter.

RELATED SPECIES: Western or **yellow salsify (*Tragopogon dubius* Scop.)** is another European species: it is less common than meadow salsify in the East and can be distinguished from it by a somewhat inflated segment of the stalk just below the flower head and the bracts subtending the flower head that are distinctly longer than the petals (ray florets).

Meadow salsify seed heads

Meadow salsify in flower along a roadside

Meadow salsify foliage

Meadow salsify flower heads

Western salsify seed heads

Western salsify flower head

Xanthium strumarium L.   Common Cocklebur

Synonyms: clotbur, rough cocklebur

Life Form: summer annual; up to 4 feet (1.2 m) tall

Place of Origin: eastern North America, Europe, and Asia

Vegetative Characteristics: The upright stems of cocklebur are well branched, hairy, and distinctly mottled with purple spots. Leaf blades, which are about 6 inches (15 cm) long, are rough to the touch and are egg- or heart-shaped with wavy, irregularly toothed margins.

Flowers and Fruit: Inconspicuous green flowers are produced from July through September. Wind-pollinated male and female flower heads occur separately on the same flower stalk (raceme)—the former above and the latter below. Fruit is a hard, egg-shaped bur approximately an inch (2.5 cm) long that contains 2 seeds. The bur is conspicuously covered with short, hooked spines that facilitate animal dispersal, and 2 sharp-pointed "beaks" project out from its tip. Burs are also buoyant and can be dispersed by water.

Germination and Regeneration: Seeds germinate in full sun and dry, well-drained soil. The larger of the 2 seeds in the bur germinates the following spring, but the smaller one typically remains dormant in the soil for several years—an interesting example of temporal as opposed to spatial seed dispersal.

Habitat Preferences: Cocklebur is particularly common along sandy ocean beaches in full sun and along river floodplains, but it also grows in a variety of disturbed urban sites including rubble dumps, vacant lots, and drainage ditches.

Ecological Functions: Disturbance-adapted colonizer of bare ground; food and habitat for wildlife; tolerance of compacted soil.

Cultural Significance: Native Americans made a tea from the leaves to treat kidney problems and diarrhea and as a blood tonic; in European herbal medicine, the root has been used to treat swollen lymph nodes and rabies. Cocklebur foliage is toxic to domesticated livestock.

Cocklebur growth habit

Spotted stems of cocklebur

Cocklebur growing in sandy coastal soil

Cocklebur in foliage

Cocklebur male (*above*) and female (*below*) flowers

Cocklebur fruits in late summer

Impatiens capensis Meerb.   Jewelweed

SYNONYMS: *Impatiens fulva, I. biflora,* spotted touch-me-not, hummingbird tree, snapweed, lady's-earrings, touch-me-not

LIFE FORM: **summer annual**; up to 6 feet (1.8 m) tall

PLACE OF ORIGIN: eastern North America

VEGETATIVE CHARACTERISTICS: Jewelweed is an extremely fast-growing plant that produces tall stems with adventitious prop roots at their base for stability. Succulent stems are smooth and hollow and exude a mucilaginous fluid when broken; they are distinctly swollen at the nodes where branches are attached and are either bluish-gray or light green depending on how much white "bloom" is on them. Alternate, oval leaves are thin, hairless, and 1–4 inches (2.5–10 cm) long with coarsely toothed margins; they are remarkable for being "unwettable" because microscopic hairs trap a thin layer of air at the leaf surface.

FLOWERS AND FRUIT: Jewelweed produces its conspicuous orange-yellow, 3-petaled flowers from June through September. Funnel-shaped blooms are approximately an inch (2.5 cm) long and have a downward-curving, nectar-filled spur at the base. They are typically pollinated by bees but are commonly visited by hummingbirds and hawkmoths. Flowers dangle freely on long stalks, and their petals are mottled with reddish-brown spots. They are followed quickly by small, green, spring-loaded capsules—about an inch (2.5 cm) long—that explode when touched, sending the seeds flying in all directions (the origin of both the scientific and common names). Peeling away the green seed coat reveals a turquoise-blue embryo, the plant's "jewel." The plant also produces small, green cleistogamous flowers that never open and are self-pollinated.

GERMINATION AND REGENERATION: Seeds germinate in early spring and grow rapidly; no regeneration occurs except from seeds. Plants die with the first frost.

HABITAT PREFERENCES: Jewelweed grows best in moist soil in either sun or shade. It is particularly common along the margins of freshwater wetlands and ponds but also grows in the understory of moist, shady woodlands. It typically forms dense stands and is one of the few native plants that can compete with garlic mustard (*Alliaria petiolata*) in the understory of second-growth forests.

ECOLOGICAL FUNCTIONS: Stream and riverbank stabilization; nutrient absorption in wetlands; food for wildlife (deer love it).

CULTURAL SIGNIFICANCE: Josselyn reported in 1672 that Native Americans used a poultice made from leaves "bruised between two stones" and mixed with "hogs grease" to treat bruises and aches. The fresh juice of the leaves has been used to reduce the swelling and inflammation caused by poison ivy and a variety of other toxic agents. Seeds are reported to be poisonous.

Jewelweed growth habit

Jewelweed illustrated in Josselyn's *New-England's Rarities* (1672)

Jewelweed flowers in profile

Hummingbird view of the jewelweed flower

Stand of jewelweed growing in a sunny wetland

Jewelweed seedpods (exploded and unexploded), green seeds, and turquoise blue embryos

Alliaria petiolata (Bieb.) Cavara & Grande   Garlic Mustard

SYNONYMS: *Alliaria officinalis, Sisymbrium alliaria,* jack-by-the-hedge, sauce-alone

LIFE FORM: winter annual or **biennial**; up to 3 feet (0.9 m) tall

PLACE OF ORIGIN: Eurasia

VEGETATIVE CHARACTERISTICS: The first-year plant is a rosette of leaves 1.25–4.5 inches (3–11.2 cm) long and wide that stays green throughout the winter and begins growing in very early spring. Kidney- or heart-shaped leaves are very distinctive, with scalloped edges, sunken veins, and overlapping bases. Second-year plants produce a tall flower stalk with alternately arranged triangular leaves. The whole plant smells of garlic when crushed, and the region where the shoot and root come together (*hypocotyl*) typically develops a distinctive S-shaped curve.

FLOWERS AND FRUIT: Garlic mustard produces small, white, 4-petaled flowers in terminal clusters in early spring. They can be either insect-pollinated or, if that fails, self-pollinated. The linear fruits that follow are 1–2 inches (2.5–5 cm) long and contain 10–20 black seeds that are forcibly ejected when the pods split open in late June or July. The entire plant dies in August, leaving behind tall, brown stalks topped with shattered seedpods.

GERMINATION AND REGENERATION: Seeds germinate in the spring or the following fall after undergoing a period of winter chilling. Seedlings can reproduce under a wide array of ecological conditions, but plants growing in moist, sunny sites produce more flowering stems and more seeds than do those growing in dry, shady sites, which are typically single-stemmed.

HABITAT PREFERENCES: Garlic mustard is a disturbance-adapted plant that can become dominant when growing in woodland understories. It is common in a variety of habitats, including shady roadside edges, forest clearings, degraded woodlands, stream and riverbanks, and landscape planting beds. Like other members of the Brassicaceae, it does not form symbiotic relationships with mycorrhizal fungi. Indeed, garlic mustard gains a competitive advantage because its roots produce chemicals that suppress the growth of the mycorrhizal fungi that inhabit the roots of other plants.

CULTURAL SIGNIFICANCE: Europeans had a long tradition of eating young garlic mustard leaves, particularly in late winter and early spring when few other greens were available; the plant also had some minor uses in folk medicine. Garlic mustard was first reported growing in North America on Long Island in 1868 but was probably introduced before then. The plant did not draw much attention until the early 1980s when it began spreading rapidly throughout the Northeast and Midwest, facilitated by the fact that deer do not eat it and by the increased levels of nitrogen in the soil as a result of increased atmospheric deposition. Many states now list garlic mustard as an invasive species.

Stand of garlic mustard

Garlic mustard rosette at the end of its first year

Garlic mustard and jewelweed dominate the understory of a Connecticut woodlot

Dead stalks of garlic mustard after two years

Garlic mustard flowers

Garlic mustard flowers and developing fruits

Garlic mustard sandwich poster from a Berlin bookshop

Barbarea vulgaris R. Br.   Yellow Rocket

SYNONYMS: St. Barbara's cress, winter cress, rocket cress, spring mustard, scurvy grass

LIFE FORM: **winter annual** or **biennial**; up to 2 feet (0.6 m) tall

PLACE OF ORIGIN: Eurasia

VEGETATIVE CHARACTERISTICS: Yellow rocket seedlings produce hairless, shiny, dark green leaves in a basal rosette that stays green through the winter. The lower leaves have a compound structure and are 2–8 inches (5–20 cm) long with a large, rounded terminal lobe and up to 5 pairs of smaller lobes lower down the petiole. In spring, as the plant bolts, it produces smaller, alternate leaves lacking lobes but with coarsely toothed margins. In bloom, yellow rocket is not as "leggy" as other yellow-flowered mustards.

FLOWERS AND FRUIT: Tall, terminal clusters (*racemes*) of bright yellow, 4-petaled flowers are produced during the spring and early summer (April to June) of their second growing season. Individual flowers are relatively small—about 0.33 inch (8 mm) across—and can be either insect- or self-pollinated. A robust plant produces hundreds of cylindrical, highly uniform fruits, each an inch or so (2.5 cm) long, which split apart at maturity to release small, oval seeds.

GERMINATION AND REGENERATION: Seeds germinate readily in spring or fall.

HABITAT PREFERENCES: Yellow rocket grows best on nutrient-rich, sandy or loamy soil. In the urban environment, it is common in minimally maintained ornamental plantings, vacant lots, abandoned grasslands, and unmowed highway banks.

CULTURAL SIGNIFICANCE: Young foliage and flowers are good to eat in the spring, tasting something like broccoli rabe. The association with St. Barbara is probably related to the fact that the greens traditionally emerged in southern Europe around the time of her feast day on December 4.

RELATED SPECIES: **Wild mustard** or **charlock (*Sinapis arvensis* L.** [formerly *Brassica kaber*]) is a European winter or summer annual that grows up to 3 feet (0.9 m) tall; its stems are covered with long, white hairs and typically show a patch of reddish pigment at the branch junctions. Leaves are alternate, smooth or slightly hairy, and up to 6 inches (15 cm) long by 2 inches (5 cm) across. Leaves low down on the stem have irregularly lobed or coarsely toothed blades with long petioles, whereas those higher up are smaller, unlobed, and clasping. Wild mustard produces dense heads (*racemes*) of pale yellow flowers at the ends of the branches between May and August. Individual flowers are insect-pollinated and are approximately 0.5 inch (1.2 cm) across with 4 non-overlapping petals. Thin fruit capsules are 1–2 inches (2.5–5 cm) long and contain numerous small, round, black seeds. The plant grows best in full sun and disturbed, rich soil and can be difficult to distinguish from a host of other "wild mustards."

Yellow rocket in full bloom

Yellow rocket flowers

Yellow rocket seedpods

Yellow rocket rosette about
to bloom in early spring

Close-up
of charlock
flowers

Charlock growth habit

Berteroa incana (L.) DC.   Hoary Alyssum

Synonyms: false hoary madwort, hoary alison

Life Form: annual, biennial, or short-lived perennial; up to 3 feet (0.9 m) tall

Place of Origin: central and eastern Europe and western Asia

Vegetative Characteristics: Hoary alyssum produces leaves that are 1.5–3 inches (3.7–7.5 cm) long by approximately 0.5 inch (1.2 cm) wide. All parts of the plant are covered with fine, star-shaped hairs that give it a distinctive grayish-green color. The plant branches from the base and develops a sprawling, multi-stemmed growth habit by the middle of summer.

Flowers and Fruit: In summer, hoary alyssum produces rounded, 1–2 inches (2.5–5 cm) wide clusters of white, insect-pollinated flowers at the tops of un-branched stems in early summer; individual flowers have 4 petals that are so deeply notched it looks like there are 8 of them. Flowers are followed by an array of plump, elliptical seedpods that line the stem. Hoary alyssum can produce flowers and fruits continuously throughout the summer and into the fall.

Germination and Regeneration: Seeds that germinate in fall produce a rosette of leaves that overwinter and flower in early summer; seeds that germinate in the spring will flower at the end of summer.

Habitat Preferences: Hoary alyssum flourishes on dry, alkaline soils with high sand or gravel content and low fertility. In the urban environment, it grows in a variety of sunny habitats with compacted soil and periodic disturbance, including walkway edges, railroad tracks, vacant lots, and compacted lawns. In the country-side, it is common along riverbanks, roadway edges, degraded agricultural fields, and worn-out pastures.

Cultural Significance: Hoary alyssum is a common weed in central and north-ern European cities and has been intentionally used to revegetate mining wastes. It is listed as a noxious agricultural weed in many U.S. states and is toxic to horses that browse its foliage. It seems that the species is becoming more common in urban and suburban areas.

Related Species: Dame's rocket (*Hesperis matronalis* L.) is a European bien-nial or short-lived perennial introduced into North America during the Colonial period for medicinal and ornamental purposes. It grows to approximately 4 feet (1.3 m) tall with alternate, lanceolate leaves up to 6 inches (15 cm) long by 2 inches (5 cm) across. It produces 4-petaled flowers in showy clusters (*racemes*) at the ends of the branches from late spring through midsummer. Individual blossoms—which are edible—are about an inch (2–2.5 cm) wide with a color that varies from purple to pink to white; they are fragrant in the evening and pollinated by butter-flies and moths. Dame's rocket grows in full sun or partial shade on moist, fertile soils and readily escapes cultivation. It is common along the disturbed edges of woodlands and flood plains and drainage ditches. Dame's rocket is often confused with garden phlox, which has opposite leaves and 5-petaled flowers that bloom later in the growing season.

Hoary alyssum growth habit

Hoary alyssum flowers and developing seedpods

Close-up of hoary alyssum flowers

Dame's rocket at the edge of a wet urban woodland

Developing dame's rocket fruits

Color variation in dame's rocket flowers

Capsella bursa-pastoris (L.) Medik.   Shepherd's Purse

SYNONYMS: shepherd's bag, pepper plant, shepherd's heart, pick-pocket

LIFE FORM: **winter annual**; up to 1–2 feet (30–60 cm) tall

PLACE OF ORIGIN: Europe

VEGETATIVE CHARACTERISTICS: Shepherd's purse seeds germinate in fall or spring to produce a basal rosette of deeply lobed leaves that are 2–4 inches (5–10 cm) long by 0.75 inch (1.8 cm) wide with a large lobe at the apex. The rosette gives rise to a tall, flowering stalk that is sparsely covered with much smaller, alternately arranged leaves. Plants produce a weak taproot and can vary greatly in size depending on growing conditions.

FLOWERS AND FRUIT: Small, inconspicuous white flowers are produced on tall, leafless stalks—up to 2 feet (60 cm) tall—in spring through midsummer. Seeds that germinate in the spring produce flowers later in the season. Individual flowers are approximately 0.12 inch (3 mm) across with 4 white petals and bloom sequentially toward the top of the raceme; they are either insect- or self-pollinated. Fruit is a heart-shaped, 2-part, flattened pod that is about 0.33 inch (8 mm) long. Reddish-brown seeds can remain viable in the soil for many years. The plant's common name derives from the shape of the seedpods.

GERMINATION AND REGENERATION: Seeds germinate readily in bare soil and disturbed sites.

HABITAT PREFERENCES: Shepherd's purse is widespread throughout the world. In an urban context, it is common in minimally maintained public parks, vacant lots, rubble dumps, minimally maintained lawns, small pavement openings and cracks, ornamental planting beds, rock outcrops, and stone walls.

CULTURAL SIGNIFICANCE: Dioscorides included shepherd's purse in his first-century herbal compendium, *De Materia Medica*. Traditional European herbalists made a tea from shepherd's purse to stop internal bleeding and hemorrhaging. The plant was an early arrival in North America and is listed in Josselyn's *New-England's Rarities Discovered* (1672) under the category: "Of such plants as have sprung up since the English planted and kept cattle in New England." Young shoots are edible in spring when cooked like spinach.

Shepherd's purse in silhouette

Dense stand of shepherd's purse

Shepherd's purse basal rosette

Mature fruits of shepherd's purse

Shepherd's purse flowers and rapidly developing fruits

Cardamine hirsuta L.   Hairy Bittercress

Synonyms: hoary bittercress, lamb's cress, land cress, shot weed

Life Form: winter or **summer annual**; up to 1 foot (30 cm) tall

Place of Origin: Eurasia

Vegetative Characteristics: Hairy bittercress seeds typically germinate in autumn and produce a basal rosette of pinnately compound leaves with 2–10 pairs of round to kidney-shaped leaflets. They are 2–4 inches (5–10 cm) long and quite conspicuous relative to the smaller stem leaves.

Flowers and Fruit: In early spring, hairy bittercress rosettes give rise to smooth flowering stalks with alternate leaves that range in height from 2–12 inches (5–30 cm). Each flower stalk is topped with a dense cluster of tiny, white flowers, each with 4 petals. Flowers are either insect- or self-pollinated and quickly give rise to slender pods approximately an inch (2.5 cm) long that explosively disperse their seeds at the slightest touch. A single vigorous plant can produce a dozen flower stalks and hundreds of seeds.

Germination and Regeneration: Hairy bittercress seeds germinate in fall and spring. The plant completes its life cycle very quickly and under cool, moist conditions is capable of producing multiple generations in a single season.

Habitat Preferences: Hairy bittercress grows best in damp, disturbed soil in either sun or shade but is also found in dry, sunny sites. It is common in cultivated garden beds where the soil is bare or has been turned over.

Cultural Significance: Leaf rosettes can be eaten in spring as a bitter herb. Left to its own devices, the plant will quickly dominate an untended vegetable garden in spring.

Related Species: Whitlow grass (*Erophila verna* (L.) DC. [formerly *Draba verna*]) is a tiny annual or biennial mustard native to Eurasia that grows to 4–5 inches (10–12.5 cm) in height. As a winter annual, it produces a basal rosette of small, lanceolate leaves—0.5–1 inch (1.2–2.5 cm) long—in late fall, which gives rise to leafless racemes of tiny white flowers approximately 0.12 inch (3 mm) across in very early spring (March or April). Its insect-pollinated flowers have 4 petals that are so deeply notched it looks like there are 8 of them. Flattened, elliptical seedpods are roughly twice as long as wide. Although whitlow grass is tiny, it is conspicuous in the landscape because it forms dense stands and blooms well before most other plants. As quickly as the plant appears, it disappears by late spring, after its seeds have been dispersed. It grows best in full sun and bare soil that is either wet or dry.

Hairy bittercress in full bloom

Hairy bittercress flowers and rapidly developing seedpods

Hairy bittercress rosette in fall

Early whitlow grass in full bloom

Hairy bittercress seedpods about to explode

Lepidium virginicum L.   Virginia Pepperweed

SYNONYMS: peppergrass, poor-man's pepper

LIFE FORM: winter or **summer annual**; up to 2 feet (60 cm) tall

PLACE OF ORIGIN: eastern North America

VEGETATIVE CHARACTERISTICS: In spring, Virginia pepperweed is a basal rosette of smooth, bright green leaves that have a large terminal lobe and numerous small lateral lobes. Later in the season, the plant produces a smooth, erect, flowering stem with small, narrow leaves approximately an inch (2.5 cm) long that do not clasp the stem.

FLOWERS AND FRUIT: Virginia pepperweed produces dense spikes of small, 4-petaled, white to greenish flowers on long, highly branched stalks (*racemes*) from June through September; they can be either insect- or self-pollinated. Flowers are followed quickly by small, slightly winged fruits with a notch at their apex. When the seedpods are mature, the plant appears to be covered with stiff bottle-brushes.

GERMINATION AND REGENERATION: Seeds germinate in fall or spring.

HABITAT PREFERENCES: This plant is ubiquitous in sunny, compacted soil, including small pavement openings, neglected residential and commercial landscapes, minimally maintained public parks, vacant lots, rubble dump sites, and unmowed highway banks and median strips.

CULTURAL SIGNIFICANCE: Darlington and Thurber (1859, 52) reported that, "This common weed is a native of the southern portion of our country, and is abundantly naturalized in many parts of Europe—thus making a partial return for the abundant supply of weeds which has crossed the ocean to our shores." Native Americans used Virginia pepperweed to treat poison ivy rash, scurvy, and the croup. Flattened pods have a strong peppery taste when eaten, as children (and some birds) like to do. Young shoots are edible in spring.

RELATED SPECIES: Field pepperweed (*Lepidium campestre* (L.) R. Br.) is a European winter annual with densely hairy, gray-green foliage and stiff, branched stems that grow to approximately 2 feet (60 cm) tall. The lower leaves are lance-shaped, up to 2 inches (5 cm) long, and taper to the base; the upper stem leaves are distinctly alternate and clasp the stem. Small, white flowers are produced on erect spikes in early summer and are followed by "bottle-brush" clusters of distinctively winged seedpods. **Field pennycress (*Thlaspi arvense* L.),** another European winter annual, is approximately 2 feet (60 cm) tall and has coarsely toothed, lance-shaped leaves that clasp the stem at their base. It produces dense racemes of small, 4-petaled, white flowers at the top of the stems from early spring through early summer; these are followed by distinctive flat, rounded pods about 0.5 inch (1.2 cm) in diameter with a notch at the apex. Young leaves are edible and taste something like mustard.

Virginia pepperweed in
flower and seed

Virginia pepperweed loaded with seed in its typical urban habitat

Stand of Virginia pepperweed in flower

Field pepperweed in seed
(note the clasping leaves)

Field pennycress in a concrete drainage pipe planter

Field pennycress
with flowers
and developing
seedpods

Humulus japonicus Sieb. & Zucc.   Japanese Hops

SYNONYMS: *Humulus scandens,* wild hops

LIFE FORM: **annual vine**; up to 20 feet (6 m) long

PLACE OF ORIGIN: temperate East Asia

VEGETATIVE CHARACTERISTICS: Stems and leaves of Japanese hops have a rough texture because of hooked bristles that catch hold of anything they touch. Dull green leaf blades have 5–7 lobes, a heart-shaped base, and are 2–5 inches (5–12.5 cm) long; on most leaves the petiole is longer than the blade. The plant has more of a sprawling than a climbing growth habit and can easily overwhelm adjacent shrubs.

FLOWERS AND FRUIT: Japanese hops produces wind-pollinated flowers in midsummer on separate male and female plants (*dioecious*) that lack petals and sepals. Male spikes are conspicuous and up to 10 inches (25 cm) long with a reddish tinge; female spikes are inconspicuous and only about an inch (2.5 cm) long. The sticky, rounded fruit clusters produced by female plants ripen in late summer.

GERMINATION AND REGENERATION: Seeds germinate readily in sunny, disturbed sites with moist soil.

HABITAT PREFERENCES: Japanese hops grows best on moist, rich soil in full sun. It is commonly found growing on river floodplains; in the urban environment, it grows on compost and rubbish piles.

ECOLOGICAL FUNCTIONS: Erosion control; food and habitat for wildlife.

CULTURAL SIGNIFICANCE: Japanese hops was introduced into North America in the late 19th century but did not attract much attention until the 1930s and 1940s, when it spread explosively throughout the East. It is much more vigorous and adaptable than the perennial European hops (*Humulus lupulus*), whose elongated fruit clusters are used to flavor beer. While several variegated varieties of Japanese hops are cultivated for ornamental purposes, none are used for flavoring beer. Several states list the plant as an invasive species.

Japanese hops foliage

Japanese hops male inflorescence

Young shoot of Japanese hops with rough hairs on the stems and leaves

Japanese hops scrambling over a pile of rubble

Mature fruit cluster of Japanese hops

Japanese hops leaf

Cerastium fontanum Baumg.   Mouseear Chickweed

SYNONYMS: *Cerastium fontanum* subsp. *vulgatum*, mouse-ear, chickweed

LIFE FORM: **herbaceous perennial**; up to 8 inches (20 cm) tall

PLACE OF ORIGIN: Eurasia

VEGETATIVE CHARACTERISTICS: Mouseear chickweed produces distinctly hairy leaves that are approximately 0.5 inch (12 mm) long by 0.25 inch (6 mm) wide; they have an opposite arrangement, lack a distinct petiole (*sessile*), and are oblong to oval. The hairy stems, which are rough to the touch, are initially upright but with time (and repeated mowing) become horizontal, forming mats up to 2 feet (60 cm) wide. Mouseear chickweed typically remains green through the winter.

FLOWERS AND FRUIT: White flowers about 0.5 inch (12 mm) wide are produced from May through October in loose clusters at the ends of the branches. Blossoms have 5 petals, each of which is deeply notched at the tip, making it look like there are twice that number; they can be either insect- or self-pollinated. Fruit is a small, cylindrical capsule full of tiny, reddish-brown seeds.

GERMINATION AND REGENERATION: Seedlings germinate throughout the year in a variety of habitats. Established plants form perennial crowns that produce new shoots that can root wherever they touch the ground, facilitating the plant's spread across the landscape. Seeds can remain dormant in the soil for many years.

HABITAT PREFERENCES: Mouseear chickweed grows best in sunny, moist soil but also tolerates dry shade. It is common in minimally maintained lawns (where it can survive mowing and form large patches), pavement openings and sidewalk cracks, vacant lots and dumps, and roadsides and median strips where its tolerance of road salt helps it survive.

ENVIRONMENTAL FUNCTIONS: Disturbance-adapted colonizer of bare ground; food and habitat for wildlife.

CULTURAL SIGNIFICANCE: Leaves and stems are edible in spring, either raw or cooked.

Close-up of mouseear chickweed flowers

Developing fruits of mouseear chickweed

Close-up of mouseear chickweed leaves

Mouseear chickweed foliage

Mouseear chickweed flowering among tall grasses by the side of a highway

Saponaria officinalis L.   Bouncing Bet

Synonyms: soapwort, old maid's pink, hedge pink, bruisewort, fuller's herb

Life Form: herbaceous perennial; up to 2 feet (60 cm) tall

Place of Origin: Eurasia

Vegetative Characteristics: Bouncing bet typically forms a cluster of unbranched stems. Leaves are smooth, opposite, dark green, 2–4 inches (5–10 cm) long by 1.5 inches (3.7 cm) wide, and oval to lance-shaped. The leaf base forms a collar around the stem, giving it a somewhat jointed appearance.

Flowers and Fruit: Bouncing bet produces terminal clusters of pale pink to cream-colored flowers from July through September. Individual flowers are very showy, are approximately an inch (2.5 cm) across, and have 5 slightly notched petals that are often bent back toward the stem (*reflexed*). Flowers are sweetly fragrant at the end of the day and are pollinated by butterflies and night-flying moths. Mutant individuals with double flowers (i.e., multiple petals) commonly grow alongside plants with normal flowers.

Germination and Regeneration: Seeds germinate readily in full sun; established plants produce new stems from short rhizomes, forming large clumps.

Habitat Preferences: Bouncing bet grows best in dry, sandy or gravelly soil and full sun. It is common along railroad tracks and in vacant lots, rubble dumps, urban meadows, roadside drainage ditches, and neglected ornamental landscapes.

Cultural Significance: One of the common names of this plant, soapwort, originated because the root, when bruised and moistened with warm water, produces a lather that can be used for washing and cleaning. The active ingredient belongs to a class of compounds known as *glycoside saponins*, which are toxic to both animals and people. The ancient Romans used the plant for washing and shrinking wool, and the Greek physician Dioscorides included it in his first-century herbal, *De Materia Medica*. European herbalists used bouncing bet externally to treat skin rashes, most notably "the itch" caused by syphilis and psoriasis; because of its toxicity its internal use was limited. Bouncing bet was intentionally introduced into North America at an early date because of its detergent and medicinal attributes. An old-fashioned, double-flowered cultivar ('Flore Pleno') is available in the nursery trade.

Bouncing bet grows well in sandy soils like those found on Cape Cod, Massachusetts

Bouncing bet foliage

Typical flowers of bouncing bet

Bouncing bet commonly produces double-flowered individuals

Bouncing bet growing along the roadside

Scleranthus annuus L.   Knawel

SYNONYM: German knotgrass

LIFE FORM: **winter annual**; up to 10 inches (25 cm) tall

PLACE OF ORIGIN: Eurasia

VEGETATIVE CHARACTERISTICS: Knawel (pronounced *nôl*) is a mat-forming plant that is often no more than a few inches (10 cm) tall. Short, linear leaves are opposite and range in length from 0.12 to 0.75 inch (3–18 mm); lateral branches arise spontaneously from the leaf axils. Under the proper conditions, knawel can form spreading clumps more than a foot (30 cm) wide with a distinctive gray-green color; it produces a small, tenacious taproot.

FLOWERS AND FRUIT: Knawel produces tiny, inconspicuous green flowers from May through October. They are difficult to see because they are sessile in the leaf axils and lack petals. Flowers are mainly self-pollinated and are quickly followed by tiny, urn-shaped fruits, each containing a single seed.

GERMINATION AND REGENERATION: Seeds germinate readily in sunny, dry locations.

HABITAT PREFERENCES: Knawel is ubiquitous along roadsides and sidewalks throughout the urban environment. It flourishes in dry soil and full sun and is tolerant of road salt. It is commonly found in the compacted soil adjacent to paved areas. The common name comes from the German word for a knot or ball of yarn.

RELATED SPECIES: Birdseye or **procumbent pearlwort** (*Sagina procumbens* L.) is a tiny, evergreen perennial that is seldom more than 2 inches (5 cm) tall by 6 inches (15 cm) wide; it grows throughout the urban environment in pavement cracks, median strips, and between walkway pavers. Birdseye pearlwort produces a basal rosette of narrow, bright green leaves—about 0.5 inch (1.2 cm) long—that give the plant a moss-like appearance. As the season progresses, it produces stolons that take root as they spread out across the ground. Flowers are inconspicuous, about 0.25 inch (6 mm) across with 4 green sepals and 4 white petals that are shorter than the sepals. Birdseye pearlwort is tolerant of roadway salt and compacted soil. It can often be found growing in full sun, but it also does well in shady, moist locations. In its native North American and Eurasian habitat, it is a cliff-dwelling species. It also grows in Antarctica where it was probably imported on the bottom of someone's boot. **Trailing pearlwort (*Sagina decumbens* (Ell.) Torr. & Gray)** is a winter or summer annual species native to North America that can grow up to 6 inches (15 cm) tall with thin, narrow leaves that are approximately 0.5 inch (1.2 cm) long. It lacks the basal rosette of evergreen leaves seen in birdseye pearlwort, and its flowers have 5 sepals and 5 white petals rather than 4 of each. It is common in the urban environment in dry, sandy soil and pavement cracks in full sun.

Knawel growth habit

Close-up of knawel foliage

Knawel thrives in compacted soil

Close-up of knawel flowers

Birdseye pearlwort grows well between walkway bricks

Birdseye pearlwort foliage is compact when trampled by people

Silene latifolia Poir.   White Campion

SYNONYMS: *Lychnis alba, Silene alba*, white cockle, evening campion, grave flower

LIFE FORM: biennial or **short-lived perennial**; up to 3 feet (0.9 m) tall

PLACE OF ORIGIN: Europe, western Asia, and northern Africa

VEGETATIVE CHARACTERISTICS: White campion initially produces a basal rosette of dull green, oval to lance-shaped leaves that are up to 5 inches (12.5 cm) long and soft-hairy to the touch. Later in the season it sends up a branched stem with opposite leaves and terminal flowers. Roots are thick and fleshy.

FLOWERS AND FRUIT: Loose clusters of white flowers, 0.75–1.5 inches (2–4 cm) wide, are produced on the end of tall stalks from May through early fall. Their most conspicuous feature is the fused sepals (*calyx*), which form a prominently ribbed, hairy "bladder" out of which emerge the 5 petals that are so deeply notched they look like 10 petals. Sweet-scented male and female flowers are produced on separate plants (*dioecious*) and open in the evening for insect-pollination. At maturity, the light brown, urn-shaped capsules release their small, brown seeds through an opening at the top that is surrounded by a crown of 10 teeth.

GERMINATION AND REGENERATION: Seeds germinate readily; established plants disturbed by cultivation can sprout from stem or root fragments.

HABITAT PREFERENCES: White campion grows best in full sun in rich, well-drained soil with a pH that is neutral or alkaline. In the urban environment, it is common along roadsides, in waste places, and in disturbed meadows.

CULTURAL SIGNIFICANCE: Like bouncing bet, the roots of white campion contain *saponins* which, when mixed with warm water, produce a lather that can be used for washing and cleaning.

RELATED SPECIES: Bladder campion or **maiden's tears** (***Silene vulgaris*** **(Moench) Garcke)** is a rhizomatous perennial with gray-green stems that grows to approximately 2 feet (60 cm) tall. It produces smooth, pale green leaves that are 2.5 inches (3.7 cm) long by 0.75 inch (1.8 cm) wide and lack a petiole. In midsummer it produces a profusion of insect-pollinated, white flowers, 0.5–0.75 inch (1.2–1.8 cm) across, that emerge from inflated, purplish or greenish bladders; they have 5 deeply notched petals that look like 10. Flowers are smaller than those of white campion, and the bladders are hairless and more rounded than elongated. Bladder campion grows best in gravelly soils in full sun and is common in vacant lots and urban meadows. The plant can apparently accumulate zinc when growing in soils contaminated with that element. Young shoots can be eaten in spring, but they are somewhat bitter.

White campion rosette foliage

Close-up of white campion flower showing inflated calyx

Bladder campion growth habit

White campion seedpod, about an inch (2.5 cm) long

Bladder campion in its typical urban habitat

Bladder campion flowers

Spergularia rubra (L.) J. & C. Presl.   Red Sandspurry

SYNONYMS: *Arenaria rubra*, roadside sandspurry, common sandspurry

LIFE FORM: **winter annual** or **short-lived perennial**; up to 6 inches (15 cm) tall

PLACE OF ORIGIN: Europe

VEGETATIVE CHARACTERISTICS: Red sandspurry may be either prostrate or upright depending on the growing conditions. Fine, narrow leaves—approximately 0.5 inch (12 mm) long—are pointed and flat with an opposite or, more typically, whorled arrangement.

FLOWERS AND FRUIT: Red sandspurry produces pinkish-red flowers in May and June and again in late summer and fall; they have 5 petals alternating with longer green sepals and are 0.5 inch (12 mm) or less in diameter. The plant attracts attention only when it is in flower and can be either self- or insect-pollinated. Fruit is a small capsule about 0.25 inch (6 mm) long that sticks up above the foliage and is filled with numerous tiny, brown seeds.

GERMINATION AND REGENERATION: Seeds germinate readily in a variety of disturbed conditions.

HABITAT PREFERENCES: Red sandspurry grows best in full sun and moist or dry soil. Because of its salt tolerance, it is common in roadside drainage ditches, compacted lawns, median strips, small pavement openings, and at the base of rock outcrops and stone walls. This inconspicuous plant grows all over the urban environment but is seldom noticed. In its native European habitat, red sandspurry grows along stream and riverbanks in sunny, gravelly or sandy soils.

CULTURAL SIGNIFICANCE: Red sandspurry has been used in traditional European medicine to treat bladder problems.

Red sandspurry growing at the Vince Lombardi Rest Area along the New Jersey Turnpike

Growth habit of red sandspurry (about 4 inches [10 cm] tall)

Red sandspurry growing in compacted soil

Red sandspurry flowers

Close-up of red sandspurry flowers

Stellaria media (L.) Vill.   Common Chickweed

SYNONYMS: *Alsine media*, starwort, starweed, winterweed, tongue-grass, chicken weed, skirt buttons, chickwhirtles

LIFE FORM: **summer** or **winter annual**; up to 1.5 feet (45 cm) tall

PLACE OF ORIGIN: Eurasia

VEGETATIVE CHARACTERISTICS: Under adverse growing conditions, chickweed is a prostrate, mat-forming plant with densely branched stems; in the absence of environmental stress the plant grows taller and lusher. Opposite leaves are broadly elliptical to egg-shaped, an inch or so (2.5 cm) long, and pointed at the tip. All parts of the plant are bright green and smooth except for a fine line of hairs that runs up one side of the stem only and then switches to the opposite side when it reaches a pair of leaves. The root system is shallow and weak.

FLOWERS AND FRUIT: Common chickweed produces small, white flowers at virtually any time of year, depending on the location; in the Northeast this is from February through November. They are approximately 0.25 inch (6 mm) in diameter and have 5 white petals that are shorter than the sepals and so deeply notched it looks like there are 10 of them. They are mainly self-pollinated. The scientific genus name of the plant, *Stellaria*, is derived from the Latin word for "star," which describes their shape. Fruit is a 1-celled capsule containing numerous tiny seeds.

GERMINATION AND REGENERATION: Seeds germinate readily in cool weather in either the fall or spring; buried seeds can remain viable for many years. Plants typically die in the heat and drought of summer. Because its stems root at the nodes, chickweed can tolerate some mowing.

HABITAT PREFERENCES: Common chickweed grows best in moist, nutrient-rich sites but is by no means limited to such areas. It is common in minimally maintained lawns, vacant lots, rubble dumps, the margins of freshwater wetlands and streams, stone walls, rock outcrops, and sidewalk cracks. Under favorable conditions, it can form large patches.

CULTURAL SIGNIFICANCE: Chickweed is a truly cosmopolitan plant that grows from high elevations in the tropics to high latitudes in Alaska and Eurasia. Young leaves and stems are eaten in rural areas throughout the world and, when lightly cooked, taste like baby spinach. The Greek physician Dioscorides included chickweed in his first-century herbal, *De Materia Medica,* and an ointment made from the plant has long been used as a folk medicine in Europe, Asia, and North America to treat skin rashes, inflammation, and bronchitis. Chickweed was an early arrival in North America and is listed in Josselyn's *New-England's Rarities Discovered* (1672) under the category: "Of such plants as have sprung up since the English planted and kept cattle in New England."

Chickweed in a typical urban niche

Chickweed growth habit

Chickweed in flower

Chickweed seedlings emerge in abundance

Chickweed can get leggy at the end of its life span

Close-up of chickweed flowers and foliage

Calystegia sepium (L.) R. Br.　Hedge Bindweed

Synonyms: *Convolvulus sepium*, wild morning glory, devil's vine, giant bindweed

Life Form: herbaceous perennial vine; stems up to 10 feet (3 m) long

Place of Origin: eastern North America and Europe

Vegetative Characteristics: Hedge bindweed produces smooth stems that trail along the ground, twining in a counterclockwise direction until they encounter something to climb on. Once it reaches the top of a plant or a fence, hedge bindweed engulfs its host with a cascade of leafy stems. Alternate leaves are smooth, 2–4 inches (5–10 cm) long and less than half as wide, with a sharp-pointed tip and squared-off basal lobes.

Flowers and Fruit: Hedge bindweed produces solitary, morning glory–like flowers from July through August in the leaf axils; petals are white or pale pink with white stripes and fused into an upward-facing, funnel-shaped tube that is 2–3 inches (5–7.5 cm) wide and similarly long. The flowers, which open in the morning and last a single day, are pollinated by bumblebees, honeybees, and hawkmoths. The 2 prominent leafy bracts that clasp the base of the flower's corolla tube are a distinctive feature. Fruit is an egg-shaped capsule containing 2–4 dark seeds that fall to the ground at maturity.

Germination and Regeneration: Seeds germinate readily in a variety of disturbed habitats; established plants produce long, fleshy rhizomes that give rise to new shoots.

Habitat Preferences: Hedge bindweed grows best in moist soils but is not limited to them. In the urban environment, it is common along the margins of wetlands and streams, stone walls, rock outcrops, ornamental planting beds, roadsides, and chain-link fences. It often grows in association with *Phragmites*.

Cultural Significance: Dioscorides included hedge bindweed in *De Materia Medica,* and the root has been used in traditional medicine as a purgative and to treat jaundice and gall bladder problems.

Related Species: **Field bindweed (*Convolvulus arvensis* L.)** is less of a climber than is hedge bindweed and commonly forms a ground-covering mat of vegetation. From June through September it produces numerous small, white, insect-pollinated flowers approximately 0.75 inch (1.8 cm) in diameter. Leaves are 1.5–2.5 inches (3.7–6.2 cm) long with pointed lobes at the base; they resemble arrowheads. Field bindweed is an herbaceous perennial that regenerates vigorously from deep-growing rhizomes and is a serious agricultural weed in many parts of the world. **Tall morning glory (*Ipomoea purpurea* (L.) Roth)** is an annual vine native to Mexico and Central America that produces alternate, smooth, heart-shaped leaves and purple, pink, blue, or white funnel-shaped flowers from midsummer into fall that are 2–3.5 inches (5–8.7 cm) wide. The plant can grow up to 10 feet (3 m) tall in good soil and commonly climbs on chain-link fences and shrubs. It is cultivated by many homeowners for its beautiful flowers. Seeds contain the psychoactive compound *ergine*.

Hedge bindweed foliage

Hedge bindweed flowers

Hedge bindweed taking over an arborvitae hedge

Field bindweed growth habit

Field bindweed flowers

Tall morning glory in flower

Sedum acre L.   Goldmoss Stonecrop

SYNONYMS: wallpepper, biting stonecrop, mossy stonecrop, goldmoss sedum

LIFE FORM: herbaceous perennial; 2–4 inches (5–10 cm) tall

PLACE OF ORIGIN: Europe, North Africa, and western Asia

VEGETATIVE CHARACTERISTICS: This low-growing, bright green ground cover stores water in its succulent, alternate leaves and can survive long periods of drought. The specific name *acre* refers to the acrid or bitter-tasting compounds in the leaves, which can also cause a skin rash in some people.

FLOWERS AND FRUIT: The plant is literally covered with bright yellow flowers in June to July; they have 5 petals, are approximately 0.4 inch (10 mm) wide, and are pollinated by a variety of insects.

GERMINATION AND REGENERATION: Goldmoss stonecrop spreads mainly by producing adventitious roots from its prostrate stems, eventually forming clumps up to 3 feet (0.9 m) in diameter.

HABITAT PREFERENCES: It flourishes in thin, dry soils in a variety of extreme natural and man-made habitats, including the upper reaches of gravelly sea shores, bare rock outcrops, dry stone walls, decaying asphalt parking lots, rubble piles, and sandy roadsides. In the northern portions of its native range, it grows near the fringes of retreating glaciers. It utilizes the specialized *Crassulacean acid metabolism* (CAM) photosynthesis found in many drought-adapted desert plants.

CULTURAL SIGNIFICANCE: Originally introduced for ornamental purposes, goldmoss stonecrop has naturalized in urban areas throughout the temperate world. It is now used extensively on green roofs.

RELATED SPECIES: Stringy or **trailing stonecrop (*Sedum sarmentosum* Bunge)** is an herbaceous perennial native to eastern Asia that produces succulent, evergreen leaves atop arching, trailing stems that root at their nodes and grow to approximately 10 inches (25 cm) long and 4 inches (10 cm) high. Flat, lance-shaped leaves are an inch or so (2.5 cm) long and arranged in whorls of 3 along the stems. Like goldmoss stonecrop, it produces bright yellow, star-shaped flowers in the summer. It is widely cultivated as an ornamental in dry, rocky soils and is readily propagated by scattering pieces of the plant on the ground. Fresh stems and leaves are edible and in Korea are a common ingredient in *bibimbap*.

Goldmoss stonecrop in a vacant lot in Connecticut

Goldmoss stonecrop (in bloom) and stringy stonecrop together on a rock

Goldmoss stonecrop in full bloom

Goldmoss stonecrop in its native habitat at the foot of a glacier in Iceland

Stringy stonecrop in a vacant lot in Detroit

Stringy stonecrop growth habit

Echinocystis lobata (Michx.) Torr. & Gray   Wild Cucumber

SYNONYMS: wild balsam apple, squirting cucumber

LIFE FORM: **annual vine**; stems more than 10 feet (3 m) long

PLACE OF ORIGIN: eastern North America

VEGETATIVE CHARACTERISTICS: Wild cucumber produces smooth, thin, green stems; leaves are alternate, smooth, star-shaped—mostly with 5 sharp-pointed lobes—and approximately 6 inches (15 cm) long. The plant climbs by means of forked tendrils that coil tightly once they find an object to attach to. The root system is remarkably spindly given that it can produce stems up to 26 feet (8 m) long. The whole plant dies with the first frost.

FLOWERS AND FRUIT: Wild cucumber is monoecious, producing separate male and female flowers on the same plant in July and August. Insect-pollinated flowers are 0.5–0.75 inch (12–18 mm) across, greenish-white in color, and have 6 narrow petals. Showy male flowers are clustered together on erect spikes 6–8 inches (15–20 cm) long; inconspicuous female flowers are produced on short spikes. Fruit, which looks like a small, spikey watermelon, is approximately 2 inches (5 cm) long by 1 inch (2.5 cm) wide.

GERMINATION AND REGENERATION: Four flat, pumpkin-like seeds are forcibly ejected from the mature fruit as it dries out, leaving behind a hollow, skeletonized shell that resembles a miniature loofah sponge dangling from the dry stalk. Seeds germinate in late spring in the shade of other plants, and seedlings rapidly climb up their stems to reach the light.

HABITAT PREFERENCES: Wild cucumber grows best when its roots are established in shady, moist soil and its foliage is exposed to full sun. In the urban environment, it is found on stockpiled topsoil and compost piles, climbing over other plants along the margins of streams and wetlands, and scrambling through and covering over ornamental plantings. In rural areas, it often completely covers stone walls. In years with good rainfall, wild cucumber can become quite rampant.

CULTURAL SIGNIFICANCE: Native Americans made a bitter tea from the root to treat stomach troubles, kidney ailments, rheumatism, and as a love potion. Charles Darwin studied and wrote about the climbing habits of this plant.

RELATED SPECIES: **Burcucumber (*Sicyos angulatus* L.)** is another annual, tendril-climbing vine native to eastern North America that can reach up to 25 feet (7.5 m) long. Alternate, medium-green leaves are approximately 8 inches (20 cm) wide and broadly 5-lobed with heart-shaped bases; they have an overall pentagon shape, as opposed to the star-shaped leaves of wild cucumber. Female and male flowers are produced separately on the same plant; both are greenish-white and not nearly as showy as those of wild cucumber. Fruits are produced in clusters on a single stalk and look like small, prickly cucumbers about 0.5 inch (12 mm) long; each fruit contains a single large seed. The plant is common along the sunny margins of streams, wetlands, and woodland thickets.

Wild cucumber smothering the surrounding vegetation

Wild cucumber female flowers

Mature wild cucumber fruit

Wild cucumber with male flowers scrambling over a clump of scouring rush

Wild cucumber climbing on a hemlock tree

Burcucumber growth habit

Burcucumber fruits

Acalypha rhomboidea Raf.   Rhombic Copperleaf

SYNONYMS: three-seeded mercury, mercury weed

LIFE FORM: **summer annual**; up to 2 feet (70 cm) tall

PLACE OF ORIGIN: eastern North America

VEGETATIVE CHARACTERISTICS: Rhombic copperleaf produces hairy to partly hairy stems. The broadly lance- or diamond-shaped (rhombic) leaves are up to 3 inches (7.5 cm) long by 1.5 inches (3.7 cm) wide, alternately arranged, and have marginal teeth. Petioles are up to 1.5 inches (3.7 cm) long, and a conspicuous bract develops where it meets the stem. Young leaves often develop a distinct copper color, and the entire plant turns coppery as it begins to die in the fall. Typically, plants develop a single taproot with weak lateral roots. Unlike other members of the Euphorbiaceae, it produces clear rather than milky sap.

FLOWERS AND FRUIT: Rhombic copperleaf produces separate male and female flowers on the same plant (*monoecious*) from June through October. Flowers are green and are located in the axils of the uppermost leaves; they are inconspicuous at the time of pollination—probably by wind—and are surrounded by a deeply notched bract with 5–9 lobes. As the female flowers develop into fruits, bracts become prominent and leaf-like, often taking on a coppery color as they reach maturity.

GERMINATION AND REGENERATION: Seeds germinate in late spring when soil temperatures start to warm up. Buried seeds can retain their viability in the soil for many years.

HABITAT PREFERENCES: Rhombic copperleaf grows best in disturbed habitats and can tolerate a wide variety of moisture, nutrient, and sunlight conditions. It is common along the unmaintained edges of sidewalks, roadways, and horticultural plantings as well as along shady stream banks and in the understory of open woodlands.

RELATED SPECIES: **Virginia copperleaf (*Acalypha virginica* L.)** is very similar to rhombic copperleaf, and the two are often confused. Bracts surrounding the flowers and fruits of Virginia copperleaf have more lobes (9–15) than those of rhombic copperleaf, and leaf petioles are typically less than half as long as the leaf blades. **Slender copperleaf (*Acalypha gracilens* A. Gray)** produces narrow leaves approximately 2 inches (5 cm) long by 0.5 inch (1.2 cm) wide, with petioles that are short relative to the length of the blade. All three species of *Acalypha* are native to eastern North America.

Rhombic copperleaf in the urban environment

Rhombic copperleaf foliage

Leafy bracts and flowers of rhombic copperleaf

Rhombic copperleaf growing as a weed in a pot of plastic flowers

Rhombic copperleaf (*left*) and slender copperleaf (*right*)

Euphorbia cyparissias L.   Cypress Spurge

SYNONYM: graveyard spurge

LIFE FORM: **herbaceous perennial**; up to a foot (30 cm) tall

PLACE OF ORIGIN: Eurasia

VEGETATIVE CHARACTERISTICS: Short, needle-like, gray-green leaves are 0.5–1.2 inches (1.2–3 cm) long by 0.12 inch (3 mm) wide and are packed tightly around the stem, giving them a distinct bottle-brush appearance. Stems and leaves exude milky sap when broken, a characteristic of the spurge family.

FLOWERS AND FRUIT: Cypress spurge produces tiny clusters of flowers subtended by conspicuous, bright yellow (occasionally pink), petal-like bracts from late spring through summer. Flowers are attractive to a wide variety of insects. Fruits are tiny capsules, each containing 3 brown seeds.

GERMINATION AND REGENERATION: Seeds germinate in sunny, dry sites but are not always produced in cold climates such as New England. Established plants reproduce strongly from root suckers and can form large colonies that are extremely difficult to eradicate.

HABITAT PREFERENCES: In the urban environment, cypress spurge is common in sunny, dry soils along railroad tracks, rock outcrops, stone walls, sidewalk edges, and neglected yards.

ECOLOGICAL FUNCTIONS: Disturbance-adapted colonizer of bare ground; erosion control on slopes; tolerant of roadway salt and compacted soil.

CULTURAL SIGNIFICANCE: Cypress spurge has been used medicinally as a laxative; its milky sap can cause dermatitis. It appears to have been introduced into North America as an ornamental ground cover sometime in the mid-1800s and has escaped cultivation to become a common inhabitant of sunny, dry habitats.

Cypress spurge in a neglected urban landscape

Cypress spurge in bloom

Cypress spurge flowers

Cypress spurge produces milky sap when injured

Cypress spurge foliage

Euphorbia maculata L.   Spotted Spurge

SYNONYMS: *Chamaesyce maculata*, *Euphorbia supina*, prostrate spurge, creeping spurge, matweed, milk purslane, spotted sandmat

LIFE FORM: **summer annual**; 1–2 inches (2.5–5 cm) tall and up to 16 inches (40 cm) wide

PLACE OF ORIGIN: eastern and central North America

VEGETATIVE CHARACTERISTICS: Spotted spurge seedlings appear late in the spring and die with the first frost. They grow pressed flat against the ground and form a dense mat with branches arranged in a near-perfect circle. Its tiny leaves are opposite, oblong to linear, and usually about 0.5 inch (1.2 cm) long and half as wide. Coloration of the leaves is variable, with most plants having a distinctive maroon blotch on the upper surface and reddish-purple undersides, but some are reddish-purple on both sides of the leaf and others lack the distinctive maroon blotch altogether. Spotted spurge stems are pink to red, densely hairy, and exude milky sap when broken. The plant produces a weak taproot, and the prostrate stems do not produce adventitious roots. Although its leaves are opposite, its branching appears to be alternate because only one bud from each leaf pair typically produces a branch.

FLOWERS AND FRUIT: Spotted spurge typically produces separate male and female flowers on the same plant (*monoecious*); they are small and inconspicuous—less than 0.12 inch (3 mm) across—and lack true sepals and petals. Flowers arise from the leaf axils in July and August and are either self- or insect-pollinated; the fruit that follows is a tiny, 3-seeded capsule.

GERMINATION AND REGENERATION: Seeds germinate readily in full sun and sandy, low-nutrient soils.

HABITAT PREFERENCES: Spotted spurge grows best in disturbed sites with sandy or gravelly soil in full sun. It tolerates both drought and soil compaction and is common in pavement openings and cracks, vacant lots, rubble dumps, trampled lawns, neglected ornamental landscapes, vegetable gardens, rock outcrops, and stone walls. As a heat-loving species, it utilizes C_4 photosynthesis.

CULTURAL SIGNIFICANCE: Native Americans used spotted spurge as a blood purifier and to treat urinary problems. The plant is toxic to livestock, and its milky sap can irritate people's skin. Spotted spurge grows throughout most of North America.

SIMILAR SPECIES: Carpetweed, purslane, and prostrate knotweed are other examples of leafy, mat-forming plants with a circular growth pattern and a distinct taproot. Of the four, only spotted spurge has milky sap.

Spotted spurge growing between a sidewalk and curb

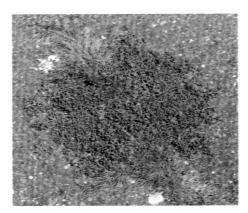

Spotted spurge in a parking lot crack

Close-up of spotted spurge flowers

Spotted spurge displays distinct alternate branching

Spotted spurge foliage

Lotus corniculatus L.   Birdsfoot Trefoil

Synonyms: cat's clover, crow toes, sheep-foot, hop o'my thumb, devil's claw, poor-man's alfalfa, eggs and bacon

Life Form: herbaceous perennial; up to 12 inches (30 cm) tall with a spread of 2 feet (60 cm)

Place of Origin: Europe and Central Asia

Vegetative Characteristics: This mat-forming plant produces alternate, nearly sessile, compound leaves with 5 small, elliptical leaflets: 3 at the tip (the source of the name trefoil) and 2 smaller ones at the base of the stalk that look like stipules. Stems trail along the ground and root at the nodes, eventually forming circular clumps that can reach up to 2 feet (60 cm) in diameter.

Flowers and Fruit: Bright yellow, pea-like flowers are produced from June through summer and into early fall; 3–10 flowers are arranged in a flat cluster at the end of a 2–4 inch (5–10 cm) stalk that puts them well above the level of the foliage. The pods that follow are about an inch (2.5 cm) long with a radial arrangement that makes them look vaguely like a bird's foot.

Germination and Regeneration: Seeds germinate readily in spring; established plants produce new shoots from a perennial crown; stolons and rhizomes develop in the fall.

Habitat Preferences: Because of its ability to "fix" atmospheric nitrogen, birdsfoot trefoil grows well in nutrient-poor, sandy soil and full sun. In the urban environment, it is common in minimally maintained lawns, vacant lots, rubble dumps, and urban meadows. It is common along highway edges and median strips because it can tolerate mowing, drought, compaction, and salt.

Ecological Functions: Soil improvement; food and habitat for wildlife; erosion control on slopes.

Cultural Significance: Seeds of this plant are a common ingredient in meadow mixes, which are widely planted for soil stabilization in difficult landscape situations. It is also cultivated as a nutritious forage crop for livestock, especially in soils that are not fertile enough for alfalfa. It was probably introduced into North America prior to 1753 as a component in hay that was accompanying the importation of cattle, as well as in ballast deposits in seaports along the Atlantic Coast.

Birdsfoot trefoil dominating a compacted lawn in early spring

Birdsfoot trefoil grows well in compacted soil

Birdsfoot trefoil inflorescence

Nitrogen-fixing root nodules of birdsfoot

Birdsfoot trefoil seedpods (the source of the plant's common name)

Medicago lupulina L.   Black Medic

SYNONYMS: trefoil, black clover, none-such, hop medic

LIFE FORM: **winter** or **summer annual**; up to 6 inches (15 cm) tall with trailing stems

PLACE OF ORIGIN: Eurasia

VEGETATIVE CHARACTERISTICS: Black medic produces spreading branches from its base. Alternate leaves consist of 3 oval to wedge-shaped leaflets, 0.4–0.8 inch (1–2 cm) long; they superficially resemble clover leaves, but the middle leaflet has a longer stalk than the lateral ones. The plant produces a shallow taproot, and the stems do not produce adventitious roots.

FLOWERS AND FRUIT: Black medic produces dense, cylindrical heads of tiny, bright yellow flowers from May through September; the flower heads, which are approximately 0.5 inch (1.2 cm) wide, are carried on short stalks and pollinated by bees. Fruit is a curved, kidney-shaped, black pod containing a single seed.

GERMINATION AND REGENERATION: Seeds germinate readily under a variety of disturbed conditions. Buried seeds retain their viability in the soil for many years.

HABITAT PREFERENCES: Black medic is highly drought tolerant and grows well on nutrient-poor soil in full sun. In the urban environment, it is common in disturbed sites with elevated soil pH, such as roadsides, compacted walkways, and waste areas. The plant's ability to "fix" atmospheric nitrogen in symbiotic association with *Rhizobium* bacteria enhances its ability to flourish in these habitats.

ECOLOGICAL FUNCTIONS: Disturbance-adapted colonizer of bare ground; soil improvement; nectar source for pollinating insects; food and habitat for wildlife.

CULTURAL SIGNIFICANCE: Black medic was introduced into North America as a forage crop for livestock. It escaped from cultivation and has become widely naturalized.

RELATED SPECIES: **Alfalfa** or **lucerne** (*Medicago sativa* L.) is a much larger perennial from south-central Asia (via Europe) that can grow up to 3 feet (0.9 m) tall and produces violet-blue flowers in flattened umbels. It is widely grown as nutritious forage for livestock (especially working horses), and the plant enriches the soil through its nitrogen-fixing root nodules. The medicinal uses of alfalfa to promote appetite and to stop bleeding are described in ancient Ayurvedic texts from India and in Dioscorides' first-century herbal, *De Materia Medica*. Its sprouted seeds are common in salad bars across North America and Europe.

Black medic
foliage and
flowers

Black medic's spreading growth habit

Black medic inflorescence

Alfalfa growth habit

Alfalfa flowers

Melilotus albus Medik.   White Sweet Clover

Synonyms: sweet or white melilot, Bokhara clover, tree clover

Life Form: biennial; up to 8 feet (2.4 m) tall

Place of Origin: Eurasia

Vegetative Characteristics: White sweet clover is a tall, widely branched plant with smooth stems. Gray-green, alternate compound leaves consist of 3 narrow leaflets, 0.5–1 inch (1.2–2.5 cm) long, with finely toothed margins; the center leaflet has a longer stalk than the lateral leaflets. One-year-old plants produce a thick taproot that stores nutrients for the plant's second-year growth spurt. The entire plant is usually dead by September, well before frost. When growing under optimal conditions, it can reach shrub-like stature.

Flowers and Fruit: White sweet clover produces upright clusters (*racemes*) of white, pea-like flowers at the ends of all its branches; flowers are approximately 0.5 inch (1.2 cm) long and are pollinated by honeybees and other insects. Small seedpods are black to gray and contain 1 or 2 seeds.

Germination and Regeneration: Seeds germinate readily in sunny, disturbed sites; they can also survive burial in the soil for many decades.

Habitat Preferences: White sweet clover grows best in full sun and dry, sandy or gravelly soil with a pH of 6.5 or higher; it is also tolerant of soils with a high clay content. It is common on roadsides, vacant lots, rubble dumps, urban meadows, unmowed highway banks and median strips, and railroad rights-of-way.

Ecological Functions: Soil improvement through nitrogen fixation; taproot breaks up compacted soil; food and habitat for wildlife.

Cultural Significance: White sweet clover was intentionally introduced from Europe as fodder for livestock, but animals that eat too much of it can experience toxic side effects because of its *coumarin* content, which also gives the plant a sweet odor. It has a long history of medicinal use in Europe to treat ulcers, inflammation, and chest complaints; to flavor cheese and tobacco; and it yields an excellent honey. The plant has been introduced as a forage plant into temperate and high-elevation subtropical zones throughout the world and now grows spontaneously in many of these areas.

Related Species: Yellow sweet clover (*Melilotus officinalis* (L.) **Pall.**) is similar to white sweet clover, except that it blooms 2–4 weeks earlier and its flowers are bright yellow rather than white.

White sweet clover growing along the roadside

White sweet clover growth habit

Nitrogen-fixing nodules on white sweet clover roots

White sweet clover flowers

Yellow sweet clover growth habit

Yellow sweet clover flowers

Securigera varia (L.) Lassen   Crownvetch

SYNONYMS: *Coronilla varia,* axseed, trailing crownvetch

LIFE FORM: **herbaceous perennial**; up to 3 feet (0.9 m) across

PLACE OF ORIGIN: Eurasia

VEGETATIVE CHARACTERISTICS: Crownvetch produces numerous sprawling stems from a long-lived, perennial base (the crown). Alternate, pinnately compound leaves are 2–6 inches (5–15 cm) long and consist of 9–25 pairs of small, oval, bristle-tipped leaflets, each approximately 0.5 inch (1.2 cm) long, and a single terminal leaflet.

FLOWERS AND FRUIT: Pea-like flowers are pink and white, about 0.5 inch (1.2 cm) long, and pleasantly fragrant; 10–25 of them are arranged in circular clusters (*umbels*) at the end of long, leafless stalks that emerge from the leaf axils. They are pollinated mainly by bees. A large plant can produce hundreds of flower clusters, making for a conspicuous display in June and early July. Flowers are followed by erect seedpods that are 1–2 inches (2.5–5 cm) long, terminated with a long beak, and subdivided into segments that each contain a hard, brown seed.

GERMINATION AND REGENERATION: Seeds germinate readily; established plants produce numerous trailing stems and rhizomes from the long-lived, perennial crown (the source of the common name).

HABITAT PREFERENCES: Crownvetch grows best in sunny, well-drained soil.

ECOLOGICAL FUNCTIONS: Tolerant of roadway salt and compacted soil; "fixes" atmospheric nitrogen in symbiosis with *Rhizobium* bacteria; food and habitat for wildlife.

CULTURAL SIGNIFICANCE: Crownvetch was introduced from China in the early 1900s by the U.S. Department of Agriculture. It was widely planted during the 1960s through the 1980s along highway embankments, especially along the newly constructed interstates. At the same time, companion plantings of crownvetch and hybrid poplar were being used for land reclamation projects in coal country. The plant has escaped cultivation in areas where it was planted and now grows spontaneously in sunny habitats with poor soil.

Crownvetch is common along highways

Compound leaves of crownvetch

Crownvetch in a vacant Detroit lot

Crownvetch flowers and foliage

Close-up of crownvetch flowers

Trifolium pratense L.   Red Clover

SYNONYMS: purple clover, meadow clover, peavine clover, cowgrass

LIFE FORM: **biennial** or **perennial**; up to 18 inches (45 cm) tall

PLACE OF ORIGIN: Eurasia

VEGETATIVE CHARACTERISTICS: Red clover is a semi-erect plant that produces numerous hairy stems from its base; leaves are finely hairy and consist of a long petiole topped with 3 elongated leaflets—up to 1.5 inches (3.7 cm) long—each typically marked with a distinctive white, V-shaped chevron on its upper surface. At night the leaflets typically droop down, probably to conserve water. Mature plants develop a tough, fibrous root system. Red clover is typically a short-lived, clump-forming plant as opposed to white clover, which is a long-lived, low-growing spreader.

FLOWERS AND FRUIT: Pink, red, or magenta flowers are arranged in dense, round clusters—about an inch (2.5 cm) across—with 1 or 2 leaves at their base. Anywhere from 50–200 individual flowers occur per head, and they are pollinated by a variety of insects (especially honeybees and bumblebees). Flowers are produced from May through October, and the typical legume seedpods develop within a month of fertilization.

GERMINATION AND REGENERATION: Red clover reproduces readily from seeds; established plants produce short rhizomes that give rise to new shoots. Its seeds, like those of other *Trifoliums*, can remain dormant in the soil for many years.

HABITAT PREFERENCES: Red clover grows in many types of soil. Its ability to "fix" atmospheric nitrogen in symbiotic association with *Rhizobium* bacteria enhances its ability to flourish in poor soils. The plant is common in abandoned grasslands, neglected residential and commercial landscapes, vacant lots, rubble dump sites, and unmowed highway banks and median strips. It is moderately tolerant of mowing, but not as tolerant as white clover.

ECOLOGICAL FUNCTIONS: Disturbance-adapted colonizer of bare ground; nectar source for pollinating insects; food and habitat for wildlife; soil improvement.

CULTURAL SIGNIFICANCE: Red clover has long been cultivated in North America as a cover crop to recondition worn-out soil and as fodder for livestock, especially when grown in combination with timothy grass. A tea made from red clover flowers was traditionally used as a cough suppressant, a mild sedative, and a blood purifier; applied externally it was used to treat ulcers, corns, and burns. Red clover is an important honey plant for bees and is the state flower of the state of Vermont.

RELATED SPECIES: **Alsike clover (*Trifolium hybridum* L.)** is an erect or ascending perennial approximately 2 feet (60 cm) tall that produces sweetly fragrant flowers in a round head at the end of a long stalk; flowers are initially white and become pink as they age, from the bottom to the top of the bicolored inflorescence. Trifoliate leaves have long petioles and lack the white markings typical of other clover species. Seeds are capable of germinating after long periods of burial in the soil.

Red clover growth habit

Red clover leaves

Red clover inflorescences

Red clover flowers

A trio of clovers: alsike (*left*), white (*center*), and red (*right*)

Alsike clover flowers

Trifolium repens L.   White Clover

Synonyms: Dutch white clover, honeysuckle clover, white trefoil

Life Form: herbaceous perennial; up to 6 inches (15 cm) tall

Place of Origin: Eurasia

Vegetative Characteristics: White clover is a mat-forming species with smooth (hairless) stems and leaves. Leaves are alternate and consist of a long petiole with 3 rounded leaflets up to an inch (2.5 cm) long; they are dark green on the upper surface, gray-green below, and are marked with a pale white, V-shaped chevron near the base. Leaflets typically droop down at night, probably to conserve water.

Flowers and Fruit: White flowers are produced throughout summer in round heads up to an inch (2.5 cm) wide on long stalks. They are pollinated by insects (especially honeybees and bumblebees) and persist on the plant in a dry state after the fruits have formed. Pods are approximately 0.25 inch (6 mm) long and contain 3–6 yellow to brown seeds.

Germination and Regeneration: White clover reproduces readily from seeds, and established plants produce stolons that spread rapidly across the surface of the ground, rooting as they go. White clover seeds are capable of germinating after long periods of burial in the soil.

Habitat Preferences: White clover is a highly adaptable plant that grows well in full sun in a variety of soil types. Because of its tolerance of mowing, white clover is a ubiquitous component of minimally maintained lawns. In the urban environment, it is common along roadway margins, rock outcrops, vacant lots, and minimally maintained meadows. The plant's ability to "fix" atmospheric nitrogen in symbiotic association with *Rhizobium* bacteria enhances its ability to flourish in these habitats. Johnson et al. (2018) have shown that urban populations of white clover are genetically distinct from rural populations.

Ecological Functions: Disturbance-adapted colonizer of bare ground; soil improvement; nectar source for pollinating insects; food and habitat for wildlife.

Cultural Significance: White clover has been used as a nutritious forage crop in North America since the 1700s, usually grown in combination with grasses. By fixing atmospheric nitrogen through its symbiotic relationship with *Rhizobium* bacteria, clover can improve the condition of low-nutrient clay or sandy soils. Numerous commercial cultivars have been selected; the most famous is 'Ladino', which is much larger than the wild type. Before broadleaf herbicides were developed, white clover was accepted as a normal component of most lawns. The plant typically produces "good luck" leaves with 4 leaflets in early spring and early fall, when day lengths are changing rapidly.

Related Species: Rabbitfoot clover (*Trifolium arvense* L.) is a European annual that can grow up to a foot (30 cm) tall; all of its parts, including the trifoliate, linear leaves, are covered with soft, reddish hairs. The plant is most conspicuous in late summer, when it produces upright clusters of pale pink to light gray flowers surrounded by silky-fringed sepals, giving them a distinctly fuzzy appearance reminiscent of a rabbit's foot. The species is extremely drought tolerant and is common along the compacted edges of roadsides and walkways.

Nitrogen-fixing nodules on white clover roots

Planting of mostly white clover

White clover spreading in a lawn

White clover inflorescence

Rabbitfoot clover growth habit

Close-up of rabbitfoot clover flowers

Vicia cracca L.   Bird Vetch

SYNONYMS: cow vetch, blue vetch, catpea, tinegrass

LIFE FORM: **herbaceous perennial vine**; 3–6 feet (0.9–1.8 m) long

PLACE OF ORIGIN: Eurasia

VEGETATIVE CHARACTERISTICS: The thin, weak stems of bird vetch either climb over other vegetation or sprawl across the ground to form a dense mat. Leaves are alternate and pinnately compound, with 8–12 pairs of linear leaflets, each 0.5–1 inch (1.2–2.5 cm) long, and a terminal tendril that facilitates climbing.

FLOWERS AND FRUIT: Individual pea-like flowers of bird vetch are insect-pollinated, deep bluish purple, and approximately 0.5 inch (1.2 cm) long. The inflorescence consists of 10–50 flowers clustered along one side of a raceme that open in sequence from the bottom to the top; racemes can reach up to 4 inches (10 cm) in length and are produced from June through August. At maturity, the flattened seedpods—up to an inch (2.5 cm) long—split into 2 twisted halves, explosively ejecting their seeds into the surrounding environment.

GERMINATION AND REGENERATION: Bird vetch reproduces readily from seed; established plants spread by rhizomes produced by the perennial crown.

HABITAT PREFERENCES: This adaptable plant does equally well in sun or shade in a variety of soil types. It is common in neglected ornamental landscapes, vacant lots, rubble dumps, urban meadows, roadside edges, and unmowed highway banks.

ECOLOGICAL FUNCTION: Disturbance-adapted colonizer of bare ground; food and habitat for wildlife; soil improvement by fixing atmospheric nitrogen.

CULTURAL SIGNIFICANCE: Originally introduced as a cover crop for worn-out soil and fodder for livestock, bird vetch has escaped cultivation to become a common component of most urban meadows. Its seeds are a popular food for birds.

Bird vetch growth habit

Bird vetch inflorescences

Bird vetch foliage with tendrils at end of its leaves

Bird vetch coming up through a yew hedge

Bird vetch flowers

Hypericum perforatum L.   Common St. Johnswort

SYNONYMS: Klamath weed, goatweed, rosin weed, tipton weed, god's wonder plant, devil's scourge

LIFE FORM: **herbaceous perennial**; up to 3 feet (0.9 m) tall

PLACE OF ORIGIN: Europe

VEGETATIVE CHARACTERISTICS: Smooth, upright stems of St. Johnswort are highly branched and somewhat woody at the base. Paired opposite leaves lack a distinct petiole and are arranged perpendicular to each other. They are elliptical to oblong, 1–1.5 inches (2.5–3.7 cm) long by 0.25–0.5 inch (6–12 mm) wide, and are covered with numerous tiny, translucent dots that are visible when the leaf is held up to the light (hence the Latin name *perforatum*).

FLOWERS AND FRUIT: Flat-topped terminal clusters of bright yellow, insect-pollinated flowers are produced from June through September. Individual flowers are 0.75–1 inch (1.8–2.5 cm) wide, and the margins of the 5 petals are marked with numerous tiny, black spots. Rubbing the flower petals between the fingers produces a reddish stain. Sharp-pointed fruit capsules contain hundreds of tiny seeds that stick to animal fur and human clothing.

GERMINATION AND REGENERATION: Seeds germinate readily but are also capable of germinating after many years of burial in the soil. Established plants produce new shoots from rhizomes that emerge from the perennial crown.

HABITAT PREFERENCES: St. Johnswort grows best in rich soil in full sun but also tolerates dry, sandy or compacted soils. In the urban environment, it is common in neglected residential and commercial landscapes, minimally maintained public parks, vacant lots, rubble dumps, abandoned grasslands, meadows, rock outcrops, stone walls, unmowed highway banks and median strips, and railroad rights-of-way.

CULTURAL SIGNIFICANCE: St. Johnswort has a long history of medicinal use in Europe. It appears in Dioscorides' first-century herbal, *De Materia Medica*, and was traditionally used as a poultice for treating wounds and ulcers. More recently it has become popular for treating mild depression along with the accompanying symptoms of fatigue, anxiety, and insomnia. It has also been used as a food preservative, especially for cheese, and as a source of reddish dye. It arrived early in North America and is listed in Josselyn's *New-England's Rarities Discovered* (1672). St. Johnswort has never been popular with farmers, mainly because it is poisonous to livestock. In 1759, John Bartram described it this way: "The common English hipericum is a very pernicious weed. It spreads over whole fields & spoils their pasturage not only by choaking ye grass but infecting our horses & sheep with scaped noses & feet especially those that have white hair on their face & legs" (p. 451). The common name *Klamath weed* originated because it was first reported as a problem in northern California near the Klamath River in the 1920s and spread rapidly to occupy some 2 million acres of rangeland by 1944. Beetles from the plant's native European habitat were introduced for biocontrol in 1945 and within ten years brought the invasion under control.

St. Johnswort
growing in
compacted soil

St. Johnswort flowers

St. Johnswort
in full bloom

St. Johnswort in August

Developing fruits of
St. Johnswort

Galeopsis tetrahit L.   Hemp Nettle

SYNONYMS: dog nettle, wild hemp, flowering nettle

LIFE FORM: **summer annual**; up to 3 feet (0.9 m) tall

PLACE OF ORIGIN: Eurasia

VEGETATIVE CHARACTERISTICS: Hemp nettle stems are conspicuously square, densely covered with bristly hairs, and form distinctive swellings just below the leaf nodes. Leaves are opposite, 1–5 inches (2.5–12.5 cm) long, and egg-shaped with a pointed tip; they have coarse teeth along their margins and long stalks. The plant branches freely from the lower leaf axils and, in rich soil, can develop a broad, spreading growth habit. Stems typically form adventitious prop roots at their base.

FLOWERS AND FRUIT: Hemp nettle produces clusters of flowers in terminal leafy spikes or in the axils of the upper leaves; the individual flowers, which are approximately 0.75 inch (1.8 cm) wide, are pale pink, or purple with magenta markings that guide insect pollinators to its nectar. The calyx that envelops the developing fruit is armed with 5 sharp, prickly spines that break off when the plant is handled. These spines and the densely hairy stems are the source of the common name. Hemp nettle dies after its seeds mature in September or October.

GERMINATION AND REGENERATION: Seeds germinate readily under a wide variety of conditions but seem to prefer moist soil in either sun or shade. A single plant can produce hundreds of seeds, which remain viable in the soil for many years. Seeds typically fall near the mother plant, and masses of seedlings often emerge the following spring.

HABITAT PREFERENCES: Hemp nettle grows best in moist soil in full sun but will tolerate dry, shady conditions. In the urban environment, it is common at the base of stone walls and rock outcrops, on the edges of minimally maintained lawns, and in disturbed waste places and vacant lots. It often forms dense stands.

CULTURAL SIGNIFICANCE: The plant has a variety of uses in European traditional medicine, and in North America the Iroquois people used an infusion of the roots to induce vomiting and as a cure for bewitchment.

RELATED SPECIES: **Motherwort (*Leonurus cardiaca* L.)** is a perennial member of the mint family with square stems and pairs of opposite leaves that are arranged perpendicular to one another (*decussate*), giving the plant a distinctive appearance. It grows best in moist, nutrient-rich soil but can also be found on dry, compacted sites in partial shade. Under optimal conditions, the plant can be 5–6 feet (1.5–1.8 m) tall with numerous side branches and a tenacious root system. The uppermost leaves that subtend the flower clusters are narrow with 3 points at their tips; the lower leaves, which resemble maple leaves, are much broader and have 3–5 distinct lobes and coarse teeth. Pink-purple, insect-pollinated flowers are produced in tight clusters in the axils of the upper leaves and occupy the top 12 inches (30 cm) or so of the plant's height. Motherwort has been used to treat a variety of "female complaints" and to treat heart palpitations at least since the time of Dioscorides, who included it in his first-century herbal, *De Materia Medica*.

Hemp nettle flowers

Hemp nettle foliage

Hemp nettle
in flower

Hemp nettle with
mature seed

Hemp nettle stem with
adventitious roots

Motherwort flowers

Glechoma hederacea L.   Ground Ivy

SYNONYMS: *Nepeta hederacea, Nepeta glechoma*, gill-over-the-ground, creeping Charlie, cats-foot, ale hoof, gillale

LIFE FORM: herbaceous perennial; stems up to 2 feet (60 cm) long

PLACE OF ORIGIN: Eurasia

VEGETATIVE CHARACTERISTICS: Ground ivy is a prostrate plant with square stems that root at the nodes. Leaves are shiny, kidney-shaped, opposite, about an inch (2.5 cm) wide, and have slightly depressed veins and broadly rounded teeth along their margins. Foliage often has a purplish cast in early spring. In mild winters, the leaves are evergreen. Under semi-shaded conditions, it can become a dominant ground cover.

FLOWERS AND FRUIT: Ground ivy produces clusters of purplish-blue flowers in the upper leaf axils of specialized stems that are shorter and more upright than the trailing vegetative stems. Funnel-shaped, insect-pollinated flowers have 2 distinct lips, are approximately 0.5 inch (1.2 cm) long, and are produced from April through July. Tiny, dry fruits are enclosed in a hairy calyx.

GERMINATION AND REGENERATION: Ground ivy will germinate from seeds, but more typically it spreads by means of its creeping stems, which root at the nodes and can form large patches.

HABITAT PREFERENCES: Ground ivy grows best in shady, moist soil but can tolerate full sun and some drought. It is common in trampled or minimally maintained lawns; neglected residential, commercial, and municipal landscapes; on the margins of freshwater wetlands, ponds, and streams; and in highway drainage ditches.

ECOLOGICAL FUNCTIONS: Disturbance-adapted colonizer of bare ground; erosion control on slopes; food and habitat for wildlife.

CULTURAL SIGNIFICANCE: Ground ivy has a multiplicity of medicinal uses in Europe that date back to at least the first century when Dioscorides included the plant in *De Materia Medica*. At one time, ground ivy was used instead of hops in fermenting and clarifying beer and ale. Traditionally, a tea made from the leaves has been used to treat diseases of the lungs and kidneys, asthma, and jaundice. The Shakers used it to treat chronic lead poisoning caused by the ingestion of paint. Ground ivy was an early arrival in North America, having been listed as a cultivated plant in Josselyn's *New-England's Rarities Discovered* (1672).

Ground ivy
in flower

Ground ivy produces long runners

Ground ivy foliage

Close-up of ground ivy flowers

Ground ivy flowering in a lawn

Lamium amplexicaule L.   Henbit

SYNONYMS: dead nettle, blind nettle, bee nettle, giraffe head

LIFE FORM: **winter annual**; up to 18 inches (45 cm) tall

PLACE OF ORIGIN: Europe

VEGETATIVE CHARACTERISTICS: Distinctive square stems of henbit can be either green or purple, and its growth habit can be either prostrate or erect. Dull green, opposite leaves are 0.5–0.75 inch (1.2–1.8 cm) long and are more or less heart-shaped with rounded teeth. The lower leaves produce a distinct petiole; the upper leaves encircle the stem with their base. Under sunny, moist conditions, the plant can form large, showy patches.

FLOWERS AND FRUIT: In early spring, henbit produces whorled clusters of showy pink to purple flowers in the axils of the uppermost leaves. The flower is a 2-lipped tube approximately 0.5 inch (1.2 cm) long that is pollinated by bees and other insects. The fruits that follow are enclosed within a small cup formed by the fused sepals.

GERMINATION AND REGENERATION: Seeds typically germinate in late summer or fall; buried seeds can remain viable in the soil for many years.

HABITAT PREFERENCES: Henbit grows best during the cool weather of spring and fall in moist, fertile soil in full sun. In the urban environment, it is common in minimally maintained lawns, ornamental planting beds, topsoil stockpiles, moist wetland edges, and roadside drainage ditches. It can form large colonies on disturbed, rich sites.

ECOLOGICAL FUNCTION: Disturbance-adapted colonizer of moist soil; food and habitat for wildlife.

CULTURAL SIGNIFICANCE: Both henbit and purple deadnettle (*below*) had limited use in traditional European medicine to stop bleeding and to promote perspiration. Young shoots are edible in spring.

RELATED SPECIES: **Purple deadnettle (*Lamium purpureum* L.)** is another European winter annual that produces light purple flowers in early spring. Unlike henbit, all of its leaves have distinct petioles; they are densely crowded together along the stem, and each pair of overlapping leaves is perpendicular to the pair of leaves immediately below or above (*decussate*), creating a pagoda-like form. The uppermost leaves of the pagoda are conspicuously purple when young, dull green as they mature, and yellow as they die. It grows best in moist rich soil, in either sun or shade.

Henbit
flowers

Henbit in flower in early spring

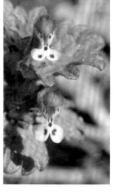

Close-up of
henbit flowers

Purple
deadnettle
dies down in
mid-to-late
spring

Purple deadnettle in flower

Conspicuous
red leaves and
flowers of purple
deadnettle

Prunella vulgaris L.   Healall

SYNONYMS: *Brunella vulgaris*, self-heal, carpenter's weed, sicle-wort

LIFE FORM: **evergreen perennial**; 2–12 inches (5–30 cm) tall

PLACE OF ORIGIN: Europe, Asia, and North America

VEGETATIVE CHARACTERISTICS: Like most mints, healall has distinctly square stems. Opposite, lance-shaped leaves have distinct petioles and are 1–4 inches (2.5–10 cm) long depending on the growing conditions. Stems creep along the ground and root at the nodes.

FLOWERS AND FRUIT: From June through September, healall produces dense, cylindrical spikes of flowers, up to 2 inches (5 cm) long, at the ends of upright stems. Insect-pollinated flowers and the bracts that make up this "head" are arranged in regular tiers or whorls; each tier is composed of a ring of 6 stalkless, light blue to purple, tubular flowers supported by a pair of spreading, sharp-pointed bracts; the number of whorls varies from 6–12. Small fruits are embedded within the leafy calyx and contain a single brown seed.

GERMINATION AND REGENERATION: Healall reproduces by seeds and by creeping stems that root at the nodes.

HABITAT PREFERENCES: Healall grows in a wide variety of disturbed habitats from sun to shade and moist to dry. It tolerates close mowing and can be abundant in minimally maintained lawns. It is also found in recently cleared woodlands, drainage ditches along highways, meadows with limestone soils, and on rock outcrops and stone walls.

ECOLOGICAL FUNCTIONS: Disturbance-adapted colonizer of bare ground; food and habitat for wildlife.

CULTURAL SIGNIFICANCE: Healall is widely distributed throughout the Northern Hemisphere. It is edible and has a long history of use as a medicinal plant wherever it is found, especially in Europe and China, where tea made from the fruiting heads has been used as a general tonic as well as to treat liver and circulation problems. In North America, the Shakers sold healall to treat hemorrhages and sore throat and to increase the flow of urine.

Growth habit and flowers of unmowed healall

Healall developing seed head

Healall going dormant in the early fall

Healall flowers in a mowed lawn

Close-up of healall flowers

Lythrum salicaria L.   Purple Loosestrife

SYNONYMS: purple lythrum, bouquet-violet, spiked loosestrife

LIFE FORM: **herbaceous perennial**; up to 5 feet (1.5 m) tall

PLACE OF ORIGIN: Eurasia

VEGETATIVE CHARACTERISTICS: Tall, square stems of purple loosestrife have lance-shaped to linear leaves that are 1–4 inches (2.5–10 cm) long and an inch (2.5 cm) wide; they are arranged opposite to one another or in whorls of 3.

FLOWERS AND FRUIT: Purple loosestrife produces terminal spikes up to 16 inches (40 cm) long with conspicuous pink-purple flowers that have 5–7 petals. The bloom period extends from July through September with the flowers at the bottom of the spike opening first and those at the top last. Following pollination by nectar-feeding insects (mainly bees), numerous small capsules develop, each containing approximately 100 small, red-brown seeds. A large plant can produce thousands of seeds, which are dispersed by flowing water or on the feet of wading birds.

GERMINATION AND REGENERATION: Purple loosestrife reproduces freely from seeds, which can remain viable in the soil for many years. Established plants produce new shoots from a perennial crown, root suckers, or detached stems, all of which contribute to the formation of large, dense stands.

HABITAT PREFERENCES: Purple loosestrife tolerates a wide range of soil moisture, pH, and nutrient conditions and can dominate the sunny margins of freshwater wetlands and meadows. It is common in brackish marshes and along the edges of ponds, streams, and rivers, especially those that have been contaminated with road salt and have a higher than normal pH. Loosestrife grows best on soil that stands just a few inches above the water table.

CULTURAL SIGNIFICANCE: Purple loosestrife was introduced into North America multiple times in the 1800s: unintentionally through ship's ballast or attached to sheep's wool, and intentionally for ornamental or medicinal purposes. It is now widespread across much of the continent. A tea made from the entire flowering plant is a European folk remedy for diarrhea and dysentery, and a wash for cleaning wounds. The wetlands where this species is dominant are a spectacular sight in July and August when the plants are in bloom. Despite its beauty, conservationists despise purple loosestrife because it outcompetes native species. Most of the states where it grows have listed it as an invasive species. Significant progress on controlling the spread of purple loosestrife has been achieved in recent years through the introduction of various insects from its native European habitat that eat the plant. Numerous horticultural varieties are in cultivation, and Charles Darwin made a detailed study of the plant's unusual floral morphology.

Purple loosestrife flowers

Purple loosestrife along the Concord River in Concord, Massachusetts

Purple loosestrife in full flower in August

Dead stems of purple loosestrife in winter

Purple loosestrife seedpods

Malva neglecta Wallr.   Common Mallow

SYNONYMS: *Malva rotundifolia*, cheeses, cheeseweed, cheese mallow, running mallow, round-leaved mallow, buttonweed

LIFE FORM: annual or **biennial**; with stems up to 3 feet (0.9 m) long

PLACE OF ORIGIN: Eurasia

VEGETATIVE CHARACTERISTICS: Under urban conditions, common mallow is a low-growing, spreading plant with numerous stems that trail along the ground with their tips turned upward. Its alternate leaves are composed of a long, grooved petiole and a roughly circular blade, anywhere from 0.75–2.5 inches (1.8–6.2 cm) wide. Leaves have wavy margins, shallow lobes, and prominent veins that radiate out from the point at which the petiole is attached. Common mallow produces a deep-growing, tenacious taproot.

FLOWERS AND FRUIT: From late spring through early fall, common mallow produces small flowers that are approximately 0.5 inch (1.2 cm) across with 5 white to pale lavender petals that are notched at the tip, often with pale violet lines along their length. Flowers emerge where the leaf is attached to the stem (the *axil*) and typically do not grow above the level of the foliage. They can be either insect- or self-pollinated. The common name "cheeses" derives from the small, flattened, circular fruit with scalloped edges, which resembles a miniature wheel of cheese.

GERMINATION AND REGENERATION: Seeds germinate in spring or fall on bare or disturbed ground, in sun or shade, and can survive burial in the soil for many years.

HABITAT PREFERENCES: Common mallow grows best in full sun and nutrient rich soil and is often found in the vicinity of penned livestock. In the urban environment, it is common in compacted lawns, neglected ornamental landscapes, vacant lots, and along walkways.

ENVIRONMENTAL FUNCTIONS: Disturbance-adapted colonizer of bare soil; food and habitat for wildlife.

CULTURAL SIGNIFICANCE: Young leaves and shoots are edible and because of their mucilaginous properties have been used to thicken soups. A tea made from the leaf or the root has been used in traditional European medicine to treat digestive problems, angina, and bronchitis. In the not too distant past, children raised on farms were fond of eating the young green fruits, or "cheeses," as a snack.

Common mallow
growing in
compacted soil

Growth
habit of
common
mallow in
flower

Common
mallow
foliage

Close-up of common mallow flower

Common mallow fruits, or "cheeses"

Mollugo verticillata L.   Carpetweed

Synonyms: Indian chickweed, devils-grip, whorled chickweed

Life Form: summer annual; up to 18 inches (45 cm) wide

Place of Origin: Central and South America

Vegetative Characteristics: The main branches of this prostrate, mat-forming plant radiate out like the spokes of a wheel and, through repeated branching, form a more or less circular mat. Narrow, light green leaves—which are widest at the tip and taper down to a narrow base—are approximately an inch (2.5 cm) long; they are produced in whorls of 3–8 at each node, separated by discrete segments of smooth stem. Carpetweed produces a slender taproot, and its prostrate branches do not root at the nodes; it utilizes a photosynthetic pathway that is intermediate between the C_3 and the C_4 types.

Flowers and Fruit: Clusters of small, greenish-white flowers with 5 petal-like sepals on slender pedicels arise in the axils of the whorled leaves from June through September. They can be either self- or insect-pollinated. Fruit is a small capsule containing numerous tiny, brown seeds.

Germination and Regeneration: Carpetweed seeds mature in the fall, and they remain dormant until late spring or early summer when the warm weather induces their germination.

Habitat Preferences: In the urban environment, carpetweed is abundant in pavement cracks and openings. Indeed, it is one of the most common weeds in parking lots. It thrives in compacted or sandy soil in full sun and is common wherever soil disturbance occurs, including ornamental planting beds, vegetable gardens, and rubble dumps.

Cultural Significance: Young plants are purported to be edible, but one would need to be very hungry.

Similar Species: Spotted spurge, purslane, and prostrate knotweed are 3 other mat-forming species with a circular growth pattern and a distinct taproot that grow in sidewalk cracks and are tolerant of being stepped on. This group of species provides a clear example of "convergent preadaptation" in urban flora.

Carpetweed growth habit

Carpetweed growing in sandy soil

Carpetweed flowers and foliage

Carpetweed surviving on air-conditioner drip in the urban environment

Carpetweed flowers

Oenothera biennis L.   Evening Primrose

SYNONYMS: scabbish, tree primrose, fever plant, evening star, night willow-herb

LIFE FORM: biennial; up to 6 feet (1.8 m) tall

PLACE OF ORIGIN: eastern and central North America

VEGETATIVE CHARACTERISTICS: The first-year plant is a basal rosette of elliptic to lance-shaped leaves, 4–8 inches (10–20 cm) long, with distinct white midveins and coarsely toothed, wavy margins; they stay green throughout the winter. Evening primrose sends up a flowering stem during its second year of growth, typically with secondary branches arising from the middle portion of the stem. Leaves become progressively smaller toward the flowering apices. It produces a fleshy, white taproot at the end of its first year.

FLOWERS AND FRUIT: Evening primrose produces conspicuous bright yellow flowers in the upper leaf axils from June through September. They are 0.75–2 inches (1.8–5 cm) in diameter with 4 large petals and fused sepals with tips reflexed downward. Flowers open toward the end of the day, emitting a fragrance that attracts primarily night-flying moths. Long, narrow seed capsules are approximately 1.5 inches (3.7 cm) long, slightly curved, and split open from the top to release numerous tiny, brown seeds; the seedpods persist conspicuously on the dead stalks through winter.

GERMINATION AND REGENERATION: Seeds typically germinate in spring in a wide variety of disturbed, sunny sites, producing a flat rosette of evergreen leaves at the end of their first year; seeds buried in the soil can retain their viability for many years and germinate when brought to the surface.

HABITAT PREFERENCES: Evening primrose grows best on dry, sandy or gravelly soil in full sun. It is common in neglected residential and commercial landscapes, minimally maintained public parks, vacant lots, rubble dumps, abandoned grasslands, meadows, small pavement openings and cracks, chain-link fence lines, unmowed highway banks and median strips, and railroad rights-of-way.

ECOLOGICAL FUNCTIONS: Disturbance-adapted colonizer of bare ground; food and habitat for wildlife.

CULTURAL SIGNIFICANCE: Evening primrose was introduced into Europe soon after North America was colonized and is now fully naturalized on that continent. Almost all parts of the plant are edible or have medicinal uses, including the roots (at the end of their first year), leaves, blossoms, flower buds and seeds, and Native American tribes made extensive use of the plant. In traditional herbal medicine, the species has been used to treat coughs, asthma, skin diseases, joint pain, and "female complaints."

Evening primrose in bloom in Boston

Evening primrose growth habit

Evening primrose inflorescence

Evening primrose growth habit under stressful seaside conditions

Evergreen evening primrose rosette at the end of its first year

Evening primrose seedpods

Oxalis stricta L.   Yellow Woodsorrel

SYNONYMS: *Oxalis europaea, Xanthoxalis stricta,* sourgrass, pickle plant

LIFE FORM: **herbaceous perennial** or **summer annual**; up to a foot (30 cm) tall

PLACE OF ORIGIN: eastern and central North America

VEGETATIVE CHARACTERISTICS: Yellow woodsorrel is a slender, erect or prostrate plant that produces weak, hairy, green or purple stems. Alternate, clover-like
leaves have long petioles and consist of 3 heart-shaped leaflets that are 0.5–1 inch
(1.2–2.5 cm) across; they fold up at night and in the heat of the noonday sun. Leaves
on most plants are green, but on some they are reddish-purple.

FLOWERS AND FRUIT: Plants produce bright yellow flowers within a month of germination and continue producing them throughout the summer. Flowers are
0.25–0.5 inch (6–12 mm) wide, have 5 petals, and are produced in clusters on long
stalks that emerge from the leaf axils. Although flowers are visited by insects, they
are mainly self-pollinated. Fruit is a 5-sided capsule that tapers to a sharp point
and is approximately 0.5 inch (1.2 cm) long.

GERMINATION AND REGENERATION: Yellow woodsorrel seedpods explode when
mature, ejecting the seeds up to 10 feet (3 m); they germinate readily the following
spring but are also capable of lying dormant in the soil for many years, waiting
for some disturbance to bring them to the surface. Established plants spread by
underground rhizomes that can grow up to 6 inches (15 cm) long.

HABITAT PREFERENCES: Yellow woodsorrel grows best in sunny locations and disturbed, nutrient-rich soils (e.g., vegetable gardens) but tolerates both shade and
drought. It is common in neglected ornamental landscapes, minimally maintained lawns, vacant lots, rubble dumps, small pavement openings, rock outcrops,
stone walls, drainage ditches, and median strips. In its native habitat, it grows in
grasslands and open woodlands.

CULTURAL SIGNIFICANCE: Yellow woodsorrel has become widely distributed
across the temperate world following its unintentional introduction into Europe
in the early 1700s. Its leaves are edible but have a sour taste due to the oxalic acid
they contain. Darlington and Thurber's (1859, 73) report that "the leaves of this
very common plant have an agreeable acidity and are frequently eaten by children" is still true today. Native American tribes used the plant medicinally to treat
a variety of ailments.

RELATED SPECIES: **Creeping woodsorrel (*Oxalis corniculata* L.)** is a southern European species similar to yellow woodsorrel in its flowers and foliage, but it has a
more prostrate growth habit and spreads by aboveground stolons that root as they
run rather than by underground rhizomes.

Yellow woodsorrel flowers, fruit, and foliage

Long white rhizomes of yellow woodsorrel

Yellow woodsorrel foliage

Yellow woodsorrel flower and foliage

Yellow woodsorrel seedpods ready to explode

Chelidonium majus L.   Greater Celandine

SYNONYMS: swallowwort, tetterwort

LIFE FORM: **biennial**; up to 3 feet (90 cm) tall

PLACE OF ORIGIN: Eurasia

VEGETATIVE CHARACTERISTICS: During its first year, greater celandine produces a rosette of pinnately compound, irregularly lobed or coarsely toothed leaves that are 3–5 inches (7.5–12.5 cm) long; they are bright green on the upper surface and whitish on the underside. When broken, the leaves and stems exude bright orange sap that dramatically stains the hands. First-year plants, which can be approximately 12 inches (30 cm) across and half as tall, are evergreen through the winter. During the second year, plants elongate to produce hollow flower stalks with alternately arranged leaves; these stems are conspicuously swollen at the nodes and easily broken. The entire plant dies in late summer or early fall of its second year.

FLOWERS AND FRUIT: Throughout the spring and summer, greater celandine produces numerous clusters of bright yellow flowers with 4 shiny petals at the ends of leafy stalks. They can be either self- or insect-pollinated. Slender, upright fruit capsules are about 1.5 inches (3.7 cm) long and explode when ripe, scattering seeds in all directions. Seeds are small and shiny with a white appendage (*elaiosome*).

GERMINATION AND REGENERATION: The white appendage attached to the seed is an attractive food source for ants who disperse the seeds when they take them back to their nest. Seeds germinate readily in early spring in a variety of disturbed sites; they can retain their viability in the soil for many years.

HABITAT PREFERENCES: Greater celandine grows best in sunny, moist, nutrient-rich soil but can tolerate shady, dry conditions. It is common in minimally maintained public parks; vacant lots and rubble dumps; the margins of freshwater wetlands, ponds, and streams; stone walls and rock outcrops; and neglected ornamental planting beds.

CULTURAL SIGNIFICANCE: Greater celandine has a long history of medicinal use. Dioscorides included it in his first-century herbal, *De Materia Medica*, and Josselyn reported in 1672 that celandine was cultivated in New England as a medicinal plant for treating various skin diseases. The orange sap, which can cause skin irritation, has been used in Europe and China to treat warts. In *American Weeds and Useful Pants* (1859, 41), Darlington and Thurber note that greater celandine is "a common weed about dwellings. Its very brittle stems, when broken, exude a saffron-colored strong-smelling juice, which is very bitter and acrid. The plant was at one time much extolled as a remedy for jaundice, but little use is made of it, except that the fresh juice is occasionally applied to warts."

Greater celandine growth habit

Greater celandine
rosette after one
growing season

Orange sap
of greater
celandine

Greater celandine growing on an old stone
bridge in Boston

Developing fruits of greater celandine

Greater celandine flowers

Phytolacca americana L. Pokeweed

SYNONYMS: *Phytolacca decandra*, poke salad, poke sallet, poke berry, poke root, skoke, pigeon berry, garget, ink berry, coakum, cancer jalap, and so on

LIFE FORM: **herbaceous perennial**; up to 6 feet (1.8 m) tall

PLACE OF ORIGIN: eastern North America

VEGETATIVE CHARACTERISTICS: Simple, alternate leaves of pokeweed are broadly lance-shaped, 4–12 inches (10–30 cm) long, and about a third as wide. They are pale green with entire margins and a smooth, waxy surface and typically wilt in the heat of the noonday sun. Under optimal conditions, pokeweed can form a large bush that dies back to a thick perennial taproot with the first frost. Fleshy, hollow stems take on a dramatic reddish-purple color in late summer and fall.

FLOWERS AND FRUIT: Pokeweed produces nodding to erect racemes of flowers from July through September. Racemes are 4–8 inches (10–20 cm) long, and the small, individual flowers have 5 white or green "petals." Flowers are mainly self-pollinated and quickly mature into long chains of deep black-purple fruits that stain the hands and that are highly attractive to both birds and children; individual fruits are approximately 0.25 inch (6 mm) across and divided into 10 segments, each of which contains a hard, black seed.

GERMINATION AND REGENERATION: Birds eagerly consume mature pokeweed fruits in the fall, and the seeds typically germinate beneath their roosts the following spring. Seeds buried in the soil for many years will germinate when brought to the surface by some form of disturbance (Del Tredici 1977). Established plants produce new stems from the perennial and highly toxic taproot, which can be up to a foot (30 cm) long by 4 inches (10 cm) wide.

HABITAT PREFERENCES: Pokeweed grows best in deep, moist soils in full sun and can live for several years under such conditions. It readily colonizes freshly disturbed ground by means of buried seeds and is common in neglected ornamental landscapes, vacant lots and rubble dumps, chain-link fence lines, unmowed highway banks, the understory of disturbed woodlands, railroad rights-of-way, and the margins of freshwater ponds and streams.

CULTURAL SIGNIFICANCE: In *American Weeds and Useful Plants* (1859, 270), Darlington and Thurber summed up pokeweed's virtues: "The young shoots of this plant afford a good substitute for Asparagus; the root is said to be actively emetic; and the tincture of the ripe berries is, or was, a popular remedy for chronic rheumatism. The mature berries have been used by the pastry cook in making pies of equivocal merit. Notwithstanding all this, the plant is regarded and treated as a weed by all neat farmers." Native American tribes used the plant medicinally and made a red dye from the juice of the berries; early settlers used the fruits to make ink. While the mature plant is considered toxic, the young shoots in spring—when less than 8 inches (20 cm) tall—are edible after boiling them for about 7 minutes. For the worried parents of young children, it should be noted that the hard, black seeds inside pokeberry fruits are toxic, but the fleshy, purple coat that surrounds them is not. Pokeweed is considered an invasive species in parts of Europe.

Pokeweed in the fall

Pokeweed has moved into this abandoned house

Pokeweed in flower

Pokeweed inflorescence

Young pokeweed plant

Maturing pokeweed fruits

Linaria vulgaris Mill.   Yellow Toadflax

SYNONYMS: butter and eggs, eggs and bacon, toadflax, wild snapdragon, flaxweed, patterns and clogs, pedlar's basket, buttered haycocks, and so on

LIFE FORM: **herbaceous perennial**; up to 3 feet (0.9 m) tall

PLACE OF ORIGIN: Europe

VEGETATIVE CHARACTERISTICS: Smooth, unbranched stems of yellow toadflax arise from underground stems to form large clumps. Narrow, linear leaves are alternate, pale gray-green, and 0.5–1.5 inches (1.2–3.7 cm) long; under stressful conditions the leaves can be so tightly packed together that the stems have a bottle-brush appearance.

FLOWERS AND FRUIT: Yellow toadflax produces compact clusters of pale yellow, snapdragon-like flowers at the ends of the stems from June through October. Individual flowers have 5 petals that are fused to form 2 lips; the lower lip has a conspicuous orange throat and an elongated spur filled with nectar. Flowers are an inch or so (2.5 cm) long and are pollinated by bees. Fruit is an egg-shaped capsule containing numerous brown seeds with a papery, circular wing.

GERMINATION AND REGENERATION: Yellow toadflax reproduces from wind-dispersed seeds; the taproot and lateral roots of established plants give rise to adventitious shoots that can form large, often circular colonies. New shoots are also produced at the base of old stems and remain green through the winter.

HABITAT PREFERENCES: Yellow toadflax tolerates drought, full sun, and compacted soil. It is extremely common in the urban environment, where it grows in vacant lots, rubble dumps, urban meadows, small pavement openings, rock outcrops and stone walls, highway banks, and railroad rights-of-way. In its native habitat, yellow toadflax is found in grasslands, sand dunes, and gravelly areas.

CULTURAL SIGNIFICANCE: Yellow toadflax has a long history of medicinal use in Europe as a purgative and a diuretic to treat jaundice and liver problems. Josselyn reported it from New England in 1672, and John Bartram described its North American "invasion" in 1759: "Ye stinking yellow linarya is ye most hurtfall plant in our pastures that can grow in our Northern climate. Neither the spade plow nor hoe can destroy it. When it is spread in a pasture every little fiber that is left will increase prodigiously. Nay some people have rolled great heaps of logs upon it & burnt them to ashes whereby ye earth was burnt half a foot deep yet it put up again as fresh as ever covering ye ground so close as not to let any grass to grow amonst it. . . . It is now spread over great part of ye inhabited parts of Pennsylvania. It was first introduced a fine garden flower but never was a plant more heartily cursed by those that suffers by its incroachment" (p. 451).

RELATED SPECIES: Oldfield toadflax (*Nuttallanthus canadensis* (L.) D.A. Sutton [formerly *Linaria canadensis*]), a native of North America, grows to approximately 2 feet (60 cm) tall and has thin, narrow leaves about an inch (2.5 cm) long. It produces numerous sky-blue flowers at the ends of tall, slender stalks from May through September. It is common in sunny, dry soil along the margins of roadsides and walkways and is quite tough despite its delicate appearance.

Yellow toadflax growth habit

Yellow toadflax foliage

Oldfield toadflax growing by a roadside

Yellow toadflax growing in gravel

Yellow toadflax flowers

Close-up of oldfield toadflax flowers

275

Plantago lanceolata L.   Buckhorn Plantain

SYNONYMS: English plantain, buckhorn, ribgrass, ribwort, black-jacks, narrow-leaved plantain, jackstraw, black jack

LIFE FORM: **herbaceous perennial**; up to 18 inches (45 cm) tall

PLACE OF ORIGIN: Europe

VEGETATIVE CHARACTERISTICS: Smooth, lance-shaped leaves have prominent parallel veins (the so-called ribs) and are 2–10 inches (5–25 cm) long and less than an inch (2.5 cm) wide. Leaves are arranged in a basal rosette with a slight twist and an erect orientation. New leaves and flowers grow up from a short underground stem that also produces a fibrous root system. In mild winters the leaves are evergreen.

FLOWERS AND FRUIT: Buckhorn plantain blooms from June through September, producing short, dense spikes of greenish-white flowers at the end of rigid, unbranched stalks. Individual flowers are wind-pollinated and open from the bottom up in a spiral pattern, often producing a distinct ring of pollen-producing stamens at any point in time. Fruit is a small capsule containing 2 seeds.

GERMINATION AND REGENERATION: Seeds become sticky when wet, which facilitates their dispersal by animals. They germinate readily in spring in sunny locations. The plant can also sprout from root pieces left in the ground following unsuccessful attempts at removal.

HABITAT PREFERENCES: Buckhorn plantain grows best in sun and tolerates compacted, pH-neutral soils and close mowing. In the urban environment, it is common in minimally maintained lawns, urban meadows, compacted pathways, vacant lots, rubble dumps, rock outcrops and stone walls, roadsides, median strips, and small pavement openings.

ECOLOGICAL FUNCTIONS: Disturbance-adapted colonizer of bare ground; food and habitat for animals.

CULTURAL SIGNIFICANCE: Buckhorn plantain has a long history of use in European folk medicine. A tea made from dry leaves is used to treat coughs, diarrhea, and dysentery, and fresh leaves can be applied directly to blisters, sores, and ulcers to reduce the pain of inflammation. In recent years it has become an important research subject for studying the complex interaction between plants and their mycorrhizal fungi.

Buckhorn plantain in an unmowed field

Buckhorn plantain dominates this neglected lawn

Buckhorn plantain growing in a crack in a wall

Buckhorn plantain foliage

Buckhorn plantain inflorescences

Plantago major L.   Broadleaf Plantain

SYNONYMS: dooryard plantain, white man's footprint, way-bread, English plantain, ribwort, ripple grass

LIFE FORM: semi-evergreen perennial; up to 12 inches (30 cm) tall

PLACE OF ORIGIN: Europe

VEGETATIVE CHARACTERISTICS: Leaves are arranged in a basal rosette and are 2–7 inches (5–17.5 cm) long and up to 4 inches (10 cm) wide. They are elliptic to egg-shaped with smooth margins, parallel veins, and a broad petiole with upward curved edges. Leaves are typically pressed flat to the ground, thereby suppressing the growth of nearby plants. Picking the leaves exposes long, stringy veins that traverse the underside of the leaf and can be removed in their entirety.

FLOWERS AND FRUIT: From June through September, broadleaf plantain produces numerous cylindrical flower stalks 4–12 inches (10–30 cm) tall that are crowded with hundreds of inconspicuous greenish-white, wind-pollinated flowers. Fruit is an oval capsule that, at maturity, splits around its midpoint when releasing its numerous seeds.

GERMINATION AND REGENERATION: Seeds germinate readily in a wide variety of habitats and soil conditions.

HABITAT PREFERENCES: While broadleaf plantain grows best in moist, nutrient-rich soil, it is remarkably tolerant of compacted, dry soils and can tolerate close mowing. In urban areas, it is common in trampled lawns, neglected ornamental landscapes, vacant lots, urban meadows, drainage ditches, and compacted pathways. In its native habitat, it grows on sunny cliffs and rocky areas.

CULTURAL SIGNIFICANCE: Native Americans reportedly called broadleaf plantain "white man's footprint" because it grew where the Europeans cleared the land. Josselyn (1672) listed the plant under the category "Of such plants as have sprung up since the English planted and kept cattle in New England," but he could well have confused it with a North American species, *Plantago rugelii* (*below*). Broadleaf plantain has long been used in Europe to treat inflammations, fevers, and sores, and—applied externally—to stop bleeding. In *Romeo and Juliet*, Shakespeare refers to this last use when Romeo tells Benvolio, "Your plantain-leaf is excellent for that" in reference to his "broken [scraped] shin." Young leaves are edible when cooked.

RELATED SPECIES: Blackseed plantain (*Plantago rugelii* Decne.) looks very similar to broadleaf plantain but is a native of North America. Seeds are black rather than brown, and its fruit capsules open by splitting below their midpoint. In general, leaves of blackseed plantain are larger, more upright, and lighter green than those of broadleaf plantain; they are also more pleated as they emerge from the basal rosette; and their petioles are reddish-purple rather than green at the bases. It is common in sunny, disturbed habitats with moist soil.

Broadleaf plantain in a typical urban niche

Stand of broadleaf plantain in full bloom

Broadleaf plantain can tolerate being driven over

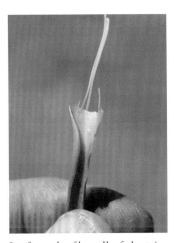

Leaf vessels of broadleaf plantain

Blackseed plantain
can dominate
moist sunny sites
that are mowed

Veronica arvensis L.   Corn Speedwell

SYNONYMS: rock speedwell, wall speedwell

LIFE FORM: winter annual; up to 4 inches (10 cm) tall

PLACE OF ORIGIN: Europe

VEGETATIVE CHARACTERISTICS: Corn speedwell can either be upright in its growth habit or form a mat with prostrate stems that radiate out from the base. Leaves are opposite, egg-shaped, 0.25–0.50 inch (6–12 mm) long, and have rounded teeth on the margins and a distinct petiole. Leaves and stems are distinctly hairy; those on the upright flowering stems are crowded together and are much smaller and narrower than the ones lower down on the stem, and they lack distinct petioles. The whole plant typically dies in summer.

FLOWERS AND FRUIT: Corn speedwell produces small, blue flowers with white centers in early-to-late spring in the upper leaf axils of the erect, leafy shoots. The tiny flowers are basically stalk-less and approximately 0.12 inch (3 mm) wide with 4 petals and 2 prominent stamens that protrude outward from the center. Flowers can be either insect- or self-pollinated, and the small, heart-shaped fruits that follow contain numerous tiny seeds.

GERMINATION AND REGENERATION: Seeds germinate mainly in fall or spring but will also sprout in summer if the weather is cool and moist. Stems root at the nodes where they touch the ground. Buried seeds of this species and most other *Veronicas* can remain viable in the soil for many years.

HABITAT PREFERENCES: Corn speedwell can grow in a wide range of soil conditions but does best in dry, sandy or gravelly soil in full sun. It is common in compacted soil in lawns and around the base of buildings, the margins of woodlands, ornamental planting beds, and small pavement openings and cracks.

CULTURAL SIGNIFICANCE: Corn speedwell and other European species have long been used in traditional European medicine as astringents to treat skin problems and as teas to treat coughs.

RELATED SPECIES: Purslane speedwell (*Veronica peregrina* L.) is a North American annual that is similar to corn speedwell in growth habit but has hairless stems, somewhat fleshy, oval leaves, and white flowers with short pedicels that are about the same size as those of corn speedwell. **Thymeleaf speedwell (*Veronica serpyllifolia* L.)** is a perennial species native to North America and Europe that produces a dense, ground-hugging mat of dark green, egg-shaped leaves with short petioles. Flowers are white with blue highlights and are carried on stalks that stand a few inches above the foliage. **Common speedwell** or **gypsyweed (*Veronica officinalis* L.)** is a Eurasian perennial that can be distinguished from other speedwells by its ability to grow well in dry, sandy soil, produce relatively large, hairy leaves up to 2 inches (5 cm) long, and spread by means of hairy stems that root as they trail along the ground. Its small, pale blue flowers with darker blue lines are carried above the foliage on spike-like racemes that are 4–6 inches (10–15 cm) tall.

Close-up of corn speedwell flower and foliage

Corn speedwell growth habit near the end of its life span; plants are about 4 inches (10 cm) tall

Purslane speedwell growth habit

Purslane speedwell flowers and foliage

Close-up of thymeleaf speedwell flower

Thymeleaf speedwell growth habit

Fallopia scandens (L.) Holub.   Climbing Buckwheat

SYNONYMS: *Polygonum scandens,* climbing false buckwheat, hedge smartweed, bindweed, winged bindweed

LIFE FORM: herbaceous perennial vine; climbing up to 15 feet (4.5 m)

PLACE OF ORIGIN: eastern North America

VEGETATIVE CHARACTERISTICS: Climbing buckwheat ascends its host by means of thin, twining stems that are wiry, hairless, and red. In the absence of a supporting plant or fence, the plant trails along the ground. Hairless, alternate leaves are 1–3 inches (2.5–7.5 cm) long and triangular to heart-shaped with 2 prominent basal lobes. A papery sheath (*ocrea*) surrounds the swollen stem at the base of each leaf.

FLOWERS AND FRUIT: Climbing buckwheat produces flowers that can be white, greenish-white, or yellowish-green from July through October. Inflorescences (*racemes*) are 2–4 inches (5–10 cm) long and emerge from the upper leaf axils, standing erect above the foliage. Individual flowers are inconspicuous and, following pollination by insects, develop rapidly into drooping stalks of green fruits, each 0.33 inch (8 mm) long with 3 distinct, papery wings.

GERMINATION AND REGENERATION: Water- or wind-dispersed seeds germinate in the moist shade of other plants; established plants resprout from a perennial crown.

HABITAT PREFERENCES: Climbing buckwheat grows well in a wide range of soil types and light conditions from full sun to full shade. In the urban environment, it is common on chain-link fences and climbing up through shrubs.

ECOLOGICAL FUNCTIONS: Disturbance-adapted colonizer of bare ground; food and habitat for wildlife.

CULTURAL SIGNIFICANCE: None known.

RELATED SPECIES: Black bindweed (*Fallopia convolvulus* (L.) A. Löve) is an annual vine from Eurasia with an appearance similar to that of climbing buckwheat, but its stems reach only approximately 5 feet (1.5 m) long, and its inflorescences are shorter—1–2 inches (2.5–5 cm) long—and have fewer flowers. Most important for identification purposes, black bindweed fruits lack conspicuous papery wings.

Climbing buckwheat in the urban environment

Conspicuous developing seeds of climbing buckwheat

Climbing buckwheat climbs by means of slender red stems

Climbing buckwheat foliage

Climbing buckwheat seeds with prominent wings

Persicaria lapathifolia (L.) Delarbre   Pale Smartweed

SYNONYMS: *Polygonum lapathifolium*, dockleaf smartweed, nodding smartweed

LIFE FORM: **summer annual**; up to 4.5 feet (1.4 m) tall

PLACE OF ORIGIN: eastern North America and Europe

VEGETATIVE CHARACTERISTICS: Pale smartweed produces multiple smooth or slightly hairy stems, either green or red, from its base. The stems are noticeably swollen and jointed at the nodes, and the papery sheath (*ocrea*) that surrounds the stem at the base of the leaf lacks a fringe of fine hairs. Purple-tinged leaves are alternate, lance-shaped to elliptic, 2–6 inches (5–15 cm) long, and are often marked with a dark splotch.

FLOWERS AND FRUIT: Pale smartweed flowers are highly variable in color—from white to pale pink to rose-colored—and are produced from July through October. They are arranged in drooping or arching terminal spikes up to 3 inches (7.5 cm) long that lack sticky hairs; small individual flowers lack petals and can either be self- or insect-pollinated. Fruits are small and black.

GERMINATION AND REGENERATION: Seeds germinate readily in sunny, disturbed sites. Buried seeds of this species are a very common "contaminant" in the topsoil used in many landscaping projects.

HABITAT PREFERENCES: Pale smartweed grows best in moist, nutrient-rich soil in full sun but also grows in dry or compacted soil. In the urban environment, it is common at the base of stone walls, in poorly drained areas of lawns and meadows, in vacant lots and waste areas, and on disturbed or compacted soil adjacent to wetlands.

ENVIRONMENTAL FUNCTIONS: Disturbance-adapted colonizer of bare ground; food and habitat for wildlife.

CULTURAL SIGNIFICANCE: Native Americans made a tea from the whole plant to treat diarrhea and made a poultice of the leaves for hemorrhoids.

RELATED SPECIES: **Pennsylvania smartweed (*Persicaria pensylvanica* (L.) M. Gómez)** is a native summer annual that resembles pale smartweed in its growth habit but produces its pale pink to rose-colored flowers in stubby, upright terminal spikes that are approximately 1.5 inches (3.7 cm) long and covered with sticky hairs. Its lance-shaped leaves are sometimes—but not always—marked in the middle with a darkly pigmented spot, and the papery sheath (*ocrea*) that surrounds the stem at the base of the leaf lacks a fringe of fine hairs. In sunny, moist conditions with good soil, the plant can grow to 4 feet (1.2 m) tall.

Pale smartweed
growth habit

Pale
smartweed
produces
a range
of flower
colors

Pale
smartweed
flowers are
typically
white

Pennsylvania smartweed growth
habit

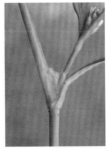

Unfringed ocrea
of Pennsylvania
smartweed

Pennsylvania smartweed
flowers

Persicaria maculosa S. F. Gray   Ladysthumb

Synonyms: *Polygonum persicaria*, heart's ease, smartweed, spotted knotweed, redshank

Life Form: summer annual; up to 3 feet (90 cm) tall

Place of Origin: Europe

Vegetative Characteristics: Ladysthumb is a densely branched plant with smooth, green or red stems that are swollen at the nodes to form distinct joints, or "knots." Leaves are lance-shaped to elliptical, 1–6 inches (2.5–15 cm) long, with a conspicuous, darkly pigmented, triangular mark in the middle of the leaf blade (the so-called lady's thumbprint). The sheath surrounding the stem "joints" (*ocrea*) is covered with short bristles. Under favorable conditions, ladysthumb will form large, spreading clumps.

Flowers and Fruit: Ladysthumb produces bright pink to red, and sometimes white, petal-less flowers from July through October. They are approximately an inch (2.5 cm) long and are arranged in dense, spike-like clusters at the ends of the branches; they can be either self- or insect-pollinated. Seeds are smooth and black.

Germination and Regeneration: Seeds germinate readily under a wide variety of conditions and can remain viable in the soil seed bank for many years.

Habitat Preferences: Ladysthumb grows best in moist soil in either sun or shade. Its size and vigor vary according to the growing conditions. It is common in neglected residential and commercial landscapes; minimally maintained public parks; vacant lots and rubble dumps; the margins of freshwater wetlands, ponds, and streams; small pavement openings and cracks; and unmowed highway banks, drainage ditches, and median strips.

Ecological Functions: Disturbance-adapted colonizer of bare ground; food for pollinating insects and foraging animals.

Cultural Significances: Ladysthumb has been used as a medicinal plant since at least the first century; Dioscorides included it in his first-century herbal, *De Materia Medica*. A tea made from the leaves was used in traditional European medicine as an astringent and to treat internal bleeding and menstrual problems. The plant was an early arrival in North America, having been listed in Josselyn's *New-England's Rarities Discovered* (1672).

Related Species: Oriental or creeping ladysthumb (*Persicaria longiseta* (Bruijn) Kitagawa (formerly *P. caespitosa*) is an annual species from Eastern Asia that can grow to 2 feet (60 cm) tall. Its general appearance is similar to *P. maculosa*, with much longer bristles on its ocreas and the pigmented spot on its leaf is not quite as dark. Oriental ladysthumb flowers from June through the end of October when it is killed by frost. It produces numerous stems from the base that take root and form large, spreading clumps. It is common along the moist, sunny edges of sidewalks and roadways as well as in the heavily shaded understory of disturbed woodlands. *P. caespitosum* was inadvertently introduced into North America in the 1920s and has been rapidly increasing in abundance since then. In some urban habitats it seems to be more common than *P. maculosa*.

Ladysthumb growth habit

Ladysthumb foliage

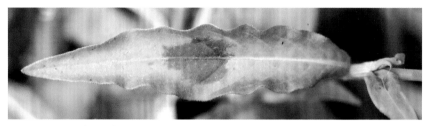

Ladysthumb leaf with "thumbprint"

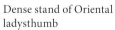

Long-bristled ocrea of
Oriental ladysthumb

Ladysthumb flowers

Dense stand of Oriental
ladysthumb

Polygonum aviculare L.   Prostrate Knotweed

Synonyms: knot-grass, door-weed, mat-grass, pink-weed, bird-grass, stone-grass, way-grass, goose-grass, allseed, centinode, nine-joints, wine's-grass

Life Form: summer annual; up to 12 inches (30 cm) tall by 24 inches (60 cm) wide

Place of Origin: Europe and northern North America

Vegetative Characteristics: Alternate, gray-green leaves are lance-shaped with pointed tips and tapered bases and are 0.5–1.3 inches (1.2–3 cm) long. Under stressful urban conditions, the plant has small leaves and forms a wiry, prostrate mat that creeps along the ground (subspecies *depressum*); in good soil it develops into a more upright plant with much larger leaves (subspecies *aviculare*) (Meerts et al. 1998). As with all *Polygonum* species, a whitish-brown, papery sheath (*ocrea*) encircles the stem at the leaf nodes or joints, giving them a "knotted" appearance. Although the plant looks like a ground cover, it does not root at the nodes, instead focusing energy on the development of a tenacious taproot. In fall, the leaves are often covered with powdery mildew.

Flowers and Fruit: Prostrate knotweed produces inconspicuous, tiny, white flowers from June though September. They are typically self-pollinated and give rise to large numbers of equally inconspicuous fruits.

Germination and Regeneration: Seeds germinate readily in early spring.

Habitat Preferences: Prostrate knotweed has a cosmopolitan distribution across the temperate world. It is ubiquitous in the urban environment and is most at home on hard, compacted soils that experience heavy pedestrian or vehicular traffic. It is common in the cracks between granite curbs and the adjacent pavement, trampled lawns (it tolerates mowing), and stone walls. In its native European habitat, prostrate knotweed grows on alkaline cliffs, the natural analog to the pavement cracks in the city. The genetics of the species are complex, and a number of polyploid races have been described (Meerts et al. 1998).

Cultural Significance: Prostrate knotweed was used in traditional European medicine to stop nosebleeds and to treat inflammations, wounds, and hemorrhoids. European races of the plant appear to have been an early arrival in North America, having been listed by Josselyn in *New-England's Rarities Discovered* (1672) under the category: "Of such plants as have sprung up since the English planted and kept cattle in New England." Darlington and Thurber reported in *American Weeds and Useful Plants* (1859, 281) that, "This humble weed is thoroughly naturalized, and is one of the commonest everywhere about dwellings." In today's urban world it has become the quintessential sidewalk crack survivor.

Similar Species: Spotted spurge, purslane, and carpetweed are other examples of mat-forming, taprooted plants with a circular growth pattern that tolerate trampling and are "preadapted" to the urban environment.

Prostrate knotweed growth form

Prostrate knotweed can grow just about anywhere

Prostrate knotweed is tolerant of compacted soil

Prostrate knotweed in its typical urban niche

Close-up of prostrate knotweed flowers

Reynoutria japonica Houtt.   Japanese Knotweed

SYNONYMS: *Fallopia japonica, Polygonum cuspidatum,* Japanese bamboo

LIFE FORM: **herbaceous perennial**; up to 8 feet (2.4 m) tall

PLACE OF ORIGIN: temperate East Asia

VEGETATIVE CHARACTERISTICS: Japanese knotweed is a fast-growing, robust perennial that can form large clumps. It is often referred to as Japanese bamboo because its stems are hollow and jointed. Large, alternate leaves are broadly egg-shaped—up to 6 inches (15 cm) long by 4 inches (10 cm) wide—with a pointed tip and a flat (*truncate*) base. A prominent papery sheath (*ocrea*) encircles the stem where leaves are attached and young leaves are rolled in the bud (*revolute*), unfurling outward from the edges as they expand. Young shoots are often suffused with red pigment; older stems are mottled with reddish spots.

FLOWERS AND FRUIT: In late summer, Japanese knotweed produces highly conspicuous clusters (*panicles*) of white flowers, up to 4 inches (10 cm) long, in the axils of its leaves. The species is *dioecious* (separate male and female plants) and pollen can be transferred by either insects or wind. Female flowers are followed by small, winged seeds that are dispersed by wind or water in late fall.

GERMINATION AND REGENERATION: Seeds germinate in a variety of conditions; established plants spread vigorously by shoots produced by deep-growing rhizomes. The plant can spread when soil containing the rhizomes is relocated.

HABITAT PREFERENCES: Japanese knotweed grows best on moist soil in full sun but tolerates shade and drought. It is salt-tolerant and common in a variety of disturbed habitats including the margins of freshwater wetlands, streams, and rivers; vacant lots and rubble dumps; and along fence lines, railroad tracks, and highway banks. It does not form a symbiotic relationship with mycorrhizal fungi.

CULTURAL SIGNIFICANCE: Japanese knotweed was originally introduced into the United States as an ornamental during the 1860s and was sold by nurseries for erosion control through the 1920s (Del Tredici 2017b). The plant has escaped cultivation, and many states (and European countries) list it as an invasive species. Once established it is notoriously difficult to eradicate. The emerging shoots, which look a bit like asparagus, are edible after boiling and have a lemony taste. The roots contain large quantities of *resveratrol*—an antioxidant also found in red wine that is reputed to promote longevity—and Japanese knotweed is cultivated in Asia to produce this "nutraceutical."

RELATED SPECIES: **Giant knotweed (*Reynoutria sachalinensis* (F. Schmidt) Nakai)** is a similar but larger growing species from northeast Asia that can reach up to 12 feet (3.6 m) in height and has leaves up to 14 inches (35 cm) long with a *cordate* base. It was introduced into North America in 1894 as a fodder plant. Japanese and giant knotweed have hybridized to produce the **Bohemian knotweed (*R.* x *bohemica* Chrtek & Chrtková)**, which, compared with Japanese knotweed, grows taller, has longer leaves with rounded bases, flowers later into the fall, and has male flower panicles that are more erect. It produces viable seed and is an invasive species in its own right.

Japanese knotweed in flower in August

Prominent ocrea and rolled leaves of Japanese knotweed

Japanese knotweed forming an arbor in the New Jersey Meadowlands

Japanese knotweed can form dense stands

Japanese knotweed in winter

Japanese knotweed fruits

Rumex acetosella L.   Red Sorrel

SYNONYMS: sheep sorrel, sourgrass, Indian cane, field sorrel, horse sorrel, sour weed, red weed, cow sorrel, sour dock

LIFE FORM: semi-evergreen perennial; up to 18 inches (45 cm) tall

PLACE OF ORIGIN: Eurasia and North Africa

VEGETATIVE CHARACTERISTICS: Red sorrel is a low-growing plant that produces a rosette of narrow, smooth leaves approximately 3 inches (7.5 cm) long and 1 inch (2.5 cm) wide with 2 prominent basal lobes that make them look like small arrowheads. Leaves have a waxy texture and produce a white ocrea where they attach to the stem. Upright flowering stalks have alternate, linear leaves that lack basal lobes. Under stressful growing conditions and at the end of the growing season, the foliage develops a deep reddish coloration. Slender, yellow roots are fibrous and freely branched.

FLOWERS AND FRUIT: Red sorrel produces loose spikes—up to 18 inches (45 cm) tall—of inconspicuous, wind-pollinated flowers from May through September. Yellow-green male flowers and reddish-brown female flowers are produced on separate plants (*dioecious*). Reddish-brown seeds mature in the fall and lack the prominent papery wings that characterize other *Rumex* species.

GERMINATION AND REGENERATION: Buried seeds can remain viable for many years and will germinate when brought to the surface by some form of disturbance (Del Tredici 1977). Established plants spread vigorously by means of rhizomes and can form large colonies, and the yellow roots produce shoots when left behind after weeding. Large patches of red sorrel that consist of only one sex are indicative of vegetative reproduction.

HABITAT PREFERENCES: Red sorrel grows best in full sun and is extremely tolerant of both acidic and poorly drained soils as well as drought. In the urban environment, it is common in neglected ornamental landscapes, compacted lawns, urban meadows, rock outcrops, and roadsides. In its native habitat, it grows in grasslands, disturbed open areas, and scree slopes.

ECOLOGICAL FUNCTIONS: Disturbance-adapted colonizer of acidic, low-fertility soils; food and habitat for wildlife.

CULTURAL SIGNIFICANCE: High concentrations of oxalic acid give red sorrel leaves a sour taste and can sicken livestock if they eat too much of it. European herbalists have used the plant as a diuretic to treat urinary and kidney problems, and in North America the Shakers sold it to treat skin diseases and as a poultice for boils and tumors. Leaves collected from plants before they flower can be eaten raw or used as a base for purees. Red sorrel is closely related to cultivated sorrel (*Rumex acetosa*).

Male clone of red sorrel in foreground with pink flowers; female clone with greenish flowers behind

Close-up of red sorrel foliage

Red sorrel growth habit

Female red sorrel in bloom

Male red sorrel in bloom

Rumex crispus L.   Curly Dock

SYNONYMS: *Rumex elongatus*, sour dock, bitter dock, yellow dock, narrow dock, dock root, garden patience

LIFE FORM: herbaceous perennial; up to 5 feet (1.5 m) tall

PLACE OF ORIGIN: Europe

VEGETATIVE CHARACTERISTICS: Curly dock starts the growing season as a rosette of shiny, narrow leaves up to a foot long (30 cm) and 2 inches (5 cm) wide with wavy, crumpled margins. In June, the flowering stems elongate with alternate leaves; a papery sheath (*ocrea*) covers the point at which the leaf attaches to the stem. The plant produces a stout, tenacious taproot.

FLOWERS AND FRUIT: Curly dock produces a tall, flowering stalk in June and July that is covered with small, greenish, wind-pollinated flowers. The fruits, which turn deep brown at maturity, consist of a single seed attached to a papery, 3-winged structure (*calyx*) that facilitates dispersal by wind or water. Distinctive, chestnut-brown seed stalks persist through the winter.

GERMINATION AND REGENERATION: Curly dock seeds germinate readily on bare ground; when buried they can remain viable for many years. Established plants resprout from a perennial crown; root fragments can give rise to new plants.

HABITAT PREFERENCES: Curly dock grows best on moist, heavy soils in full sun. In the urban environment, it is common in neglected ornamental landscapes, vacant lots, rubble dumps, urban meadows, unmowed highway banks, drainage ditches, and railroad rights-of-way.

CULTURAL SIGNIFICANCE: The root has a long history of use in European folk medicine as a laxative; an astringent; and a treatment for lung, liver, and skin problems. Young, unfurled leaves that emerge from the center of the plant in spring are edible after blanching. Curly dock was an early arrival in North America, having been listed in Josselyn's *New-England's Rarities Discovered* (1672) under the category: "Of such plants as have sprung up since the English planted and kept cattle in New England."

RELATED SPECIES: Broadleaf dock (*Rumex obtusifolius* L.) is a European perennial that grows to approximately 4 feet (1.2 m) tall. Its leaves are flatter and much broader than those of curly dock; they are heart-shaped at the base and lack a wavy margin. Like curly dock, broadleaf dock grows in nutrient-rich, sunny locations but is more shade tolerant and needs a bit less moisture. According to Darlington and Thurber (1859, 284), "This species is even more worthless than the preceding [*R. crispus*]. . . . The presence of either imparts a very slovenly appearance to a meadow or pasture lot." Broadleaf dock was once a popular treatment for nettle stings.

Curly dock in
Tuscany, Italy

Persistent seed heads of curly dock

Curly dock by the side of a Detroit street

Broadleaf dock growth habit

Broadleaf dock foliage

Portulaca oleracea L.   Common Purslane

SYNONYMS: pussley, pursley, pressley, wild portulaca, little hogweed

LIFE FORM: **summer annual**; up to 6 inches (15 cm) tall by 2 feet (60 cm) wide

PLACE OF ORIGIN: a cosmopolitan plant of Eurasian origin

VEGETATIVE CHARACTERISTICS: Common purslane is a prostrate, mat-forming plant with smooth, red or green stems that grow out to form a nearly perfect circle. Succulent, alternate leaves are approximately an inch (2.5 cm) long and are shaped like a wedge or a paddle—broad at the tip and narrow at the base. Leaves have a smooth, glossy surface that makes them glisten in bright sunlight. The plant develops a distinct taproot, and its stems typically do not root at the nodes.

FLOWERS AND FRUIT: Purslane starts producing bright yellow flowers in July, roughly a month after germination, and continues producing them through September. Flowers have 5 short petals, lack a distinct stalk or petiole, and are clustered at the ends of the branches or in the leaf axils. They typically open only when it is sunny, and they are mostly self-pollinated. Small, globe-shaped capsules are approximately 0.25 inch (6 mm) long and contain numerous tiny, black seeds.

GERMINATION AND REGENERATION: Purslane is a "hot-weather weed" whose seeds typically do not begin germinating until the soil warms up in June. Seeds will continue sprouting throughout the summer, especially after rain. Plants that have been uprooted and discarded on the soil surface can use the water stored in their succulent stems and leaves to reorient their root system and continue growing. Buried seeds can retain their viability in the soil for up to 40 years.

HABITAT PREFERENCES: Purslane grows best in sandy, nutrient-rich soils in full sun but tolerates dry, compacted soil. It is common in neglected residential and commercial landscapes, vacant lots, rubble dumps, compacted or worn-out lawns, horticultural planting beds, and small pavement openings and cracks. It is a heat-loving species that utilizes C_4 photosynthesis.

CULTURAL SIGNIFICANCE: Purslane is widely distributed throughout the temperate and subtropical regions of the world. The species name, *oleracea*, means "of the vegetable garden" and refers to the plant's edibility. Wherever purslane grows—either wild or cultivated—people have eaten its mucilaginous leaves and stems raw or used them to thicken soups and stews. Leaves have a sharp, acidic taste and are a rich source of omega-3 fatty acids. The plant has a long history of medicinal use in Europe and Asia, mainly to reduce inflammation. Dioscorides included it in his first-century herbal, *De Materia Medica*. Purslane was being cultivated as early as 1631 at John Winthrop's Plymouth Colony in New England. Horticultural selections with leaves that are larger than normal or that are reddish or yellowish are available in the nursery trade.

SIMILAR SPECIES: Spotted spurge, carpetweed, and prostrate knotweed are other examples of mat-forming plants with a circular growth pattern and a distinct taproot that are common in the urban environment. Of the four, purslane is the only succulent species.

Purslane growth habit

Purslane dominating a patch of meadow

Purslane in its typical urban niche

Purslane foliage and flowers

Growth habit of a purslane seedling

Ranunculus bulbosus L.   Bulbous Buttercup

SYNONYMS: bulbous crowfoot, yellowweed, blister flower, gowan

LIFE FORM: herbaceous perennial; up to 18 inches (45 cm) tall

PLACE OF ORIGIN: Eurasia

VEGETATIVE CHARACTERISTICS: Bulbous buttercup is a relatively low-growing plant with a basal rosette of 3-parted compound leaves; the middle leaflet has a distinct stalk and the 2 lateral leaflets are stalkless (*sessile*); the basal leaves have long, hairy petioles. The plant produces a distinct bulb approximately 0.5 inch (1.2 cm) wide at its base.

FLOWERS AND FRUIT: Iconic, shiny yellow flowers are produced at the ends of sparsely leaved, hairy stems from April through June. Insect-pollinated flowers are approximately an inch (2.5 cm) wide with 5–7 petals that are rounded at the apex and wedge-shaped at the base. Green sepals below the petals curve back against the stem (*recurved*). Each flower gives rise to a globe-shaped cluster of flattened, dry fruits.

GERMINATION AND REGENERATION: Seeds germinate readily in spring or fall in sunny sites; established plants sprout from a perennial crown.

HABITAT PREFERENCES: Bulbous buttercup is common in dry lawns; neglected residential and commercial landscapes; minimally maintained public parks and open spaces; vacant lots and rubble dump sites; freshwater wetlands and streams; rock outcrops and stone walls; and unmowed highway banks and median strips with frequent salt applications.

CULTURAL SIGNIFICANCE: Leaves and stems of this plant and of *R. acris* (see below) contain toxic, acrid juices that discourage cattle and horses from eating them, and the plants often become dominant in overgrazed pastures. Bulbous buttercup has had limited use as a treatment for arthritis and rheumatism because of its caustic properties. Bulbs are supposedly edible when collected in early spring and thoroughly dried. In the familiar childhood game, the reflected glow of the flower's yellow petals held under the chin identifies one as a lover of butter.

RELATED SPECIES: Tall buttercup (*Ranunculus acris* L.) is another species from Eurasia; it is taller than bulbous buttercup—reaching up to 3 feet (0.9 m)—and lacks the distinct bulb at the base of the stem. Its large, hairy leaves are deeply divided into 3–7 lobes; the central lobe lacks a distinct stalk. Flowers of the two species are similar, but the sepals of tall buttercup are relatively small and inconspicuous and are not recurved. Tall buttercup is not as drought tolerant as bulbous buttercup. Because livestock shun the plant, it often becomes dominant in overgrazed pastures.

Bulbous buttercup bulb

Bulbous buttercup growth habit and flowers

Bulbous buttercup flower

Tall buttercup flower (note non-recurved sepals) and seed head

Tall buttercup growth habit

Tall buttercup flowers

Ranunculus repens L.   Creeping Buttercup

SYNONYMS: spotted buttercup, creeping crowfoot

LIFE FORM: **herbaceous perennial**; up to a foot (30 cm) tall

PLACE OF ORIGIN: Eurasia

VEGETATIVE CHARACTERISTICS: Creeping buttercup is typically a low-growing plant with leaves arranged in a basal rosette. Dark green leaves are often mottled with faint white markings; they are roughly 4 inches (10 cm) long and wide and consist of 3 distinct segments, the middle one of which is distinctly stalked. Margins of all the leaflets are irregularly lobed.

FLOWERS AND FRUIT: Creeping buttercup produces shiny, bright yellow, insect-pollinated flowers from May through July that are approximately an inch (2.5 cm) across and have 5–7 petals; the sepals are not reflexed back against the stem. Tiny, spherical fruits have a curved beak.

GERMINATION AND REGENERATION: Seeds germinate readily in moist soils; established plants produce runners (stolons) that creep along the ground, rooting as they go and forming large, spreading patches.

HABITAT PREFERENCES: Creeping buttercup grows best in moist, humus-rich soil but also thrives in poorly drained sandy or gravelly soil; it seems to grow equally well in sun or shade. In its native habitat, creeping buttercup grows in grasslands and along stream banks.

CULTURAL SIGNIFICANCE: Creeping buttercup was once cultivated as an ornamental ground cover; its foliage, when fresh, is toxic to livestock.

RELATED SPECIES: **Lesser celandine, pilewort, or fig-buttercup (*Ficaria verna* Huds.** [formerly *Ranunculus ficaria* L.]) is a clump-forming Eurasian species with glossy, heart- or kidney-shaped leaves that are up to 2 inches (5 cm) long and wide, with wavy, toothed edges. It leafs out very early in spring and produces solitary, bright yellow flowers in April that stand 2–6 inches (5–15 cm) above the foliage. The plant grows best in full sun to light shade in moist, loamy soils but can tolerate drier, shadier conditions. Most of the plants growing in North America are derived from showy, tetraploid varieties (subspecies *bulbifera*) introduced from Europe for ornamental purposes at an unknown date. Flowers are mostly sterile and the plant reproduces exclusively from fleshy tubers located at the base of the plant and from bulbils produced in the upper leaf axils. Since the 1980s, lesser celandine has been spreading aggressively into wetlands, flood plains, and moist roadsides where it often becomes dominant by virtue of its ability to reproduce vegetatively; it is listed as an invasive species in many states. Lesser celandine has been used in traditional European medicine to treat hemorrhoids (piles) and is toxic to livestock.

Creeping buttercup in flower

Creeping buttercup foliage and flowers

Lesser celandine flowers

A large patch of lesser celandine in a wet meadow

Lesser celandine growth habit

Lesser celandine foliage

Potentilla argentea L.   Silvery Cinquefoil

SYNONYM: silver five-fingers

LIFE FORM: **herbaceous perennial**; up to 1.5 feet (45 cm) tall

PLACE OF ORIGIN: Eurasia

VEGETATIVE CHARACTERISTICS: Silver cinquefoil can grow as an erect plant when left undisturbed or as a prostrate mat when mowed. Alternate, palmately compound leaves consist of 5–7 coarsely toothed leaflets about an inch (2.5 cm) long; their upper surfaces are dark green and glossy, and their undersides are covered with a dense mat of silvery white hairs (the source of the common name), which makes the plant easy to identify.

FLOWERS AND FRUIT: From late spring through the summer, silver cinquefoil produces small, yellow, insect-pollinated flowers that have 5 petals and are approximately 0.5 inch (1.2 cm) in diameter. In May and June, the plant can transform compacted lawns into a sea of yellow. Small, dry fruits (*achenes*) are clustered together on a hairy receptacle. The plant also has the capacity to produce seeds asexually (*apomixes*).

GERMINATION AND REGENERATION: Seeds germinate readily on dry, sunny sites and are capable of germinating after a long period of burial in the soil. Established plants produce new shoots from a deep-rooted perennial crown.

HABITAT PREFERENCES: Silver cinquefoil grows best in full sun and dry, compacted soils with low fertility—including worn-out lawns.

CULTURAL SIGNIFICANCE: In traditional medicine, the root is considered an astringent and has a variety of minor uses.

RELATED SPECIES: **Oldfield cinquefoil (*Potentilla simplex* Michx.),** also known as "five fingers," is a low-growing native species with palmately compound leaves that have 3–5 coarsely toothed leaflets. Wiry, red or green runners (*stolons*) grow out from established plants, strawberry style, to produce new plantlets. In cold areas, the plant typically goes underground in winter, but in mild climates it is evergreen. The plant produces bright yellow flowers singly on long stalks from May through June; they are 0.5–0.75 inch (1.2–1.8 cm) wide and have 5 broad petals. Small, dry fruits (*achenes*) are clustered together on a hairy receptacle. The plant grows best in sunny locations and is an indicator of sandy soils that are acidic and low in nutrients.

Silver cinquefoil in flower produces the yellow haze in this park

Silver cinquefoil foliage

Silver cinquefoil in full flower

Oldfield cinquefoil habit and foliage (note the long runners)

Oldfield cinquefoil flower

Potentilla recta L.   Sulfur Cinquefoil

SYNONYMS: sulfur five-fingers, upright cinquefoil, rough-fruited cinquefoil

LIFE FORM: **herbaceous perennial**; up to 30 inches (75 cm) tall

PLACE OF ORIGIN: Eurasia

VEGETATIVE CHARACTERISTICS: Sulfur cinquefoil stems are hairy and upright; the hairy, palmately compound leaves consist of 5–7 lance-shaped leaflets—up to 2 inches (5 cm) long—with coarsely toothed margins and a long petiole. The upper leaves are smaller and have only 3 leaflets and a short or no petiole. New shoots emerge from a woody rootstock.

FLOWERS AND FRUIT: Showy, pale yellow flowers are produced from June through August. The flowers—which are arranged in flat clusters at the ends of the branches—are approximately an inch (2.5 cm) in diameter and have 5 petals that are shallowly notched at the tip; they can be either insect- or self-pollinated. Dry fruits (*achenes*) are clustered together on a hairy receptacle.

GERMINATION AND REGENERATION: Seeds germinate readily in sunny, dry sites and can survive burial in the soil for many years. Established plants produce new shoots from dormant buds located at the periphery of the old stem, eventually forming a circular cluster of plants.

HABITAT PREFERENCES: Sulfur cinquefoil grows well in sunny, dry locations. In the urban environment, it is common in waste areas, dumps, unmowed highway banks, urban meadows, and roadside ditches.

CULTURAL SIGNIFICANCE: The pale yellow flowers of sulfur cinquefoil are quite beautiful, but the plant has the capacity to spread aggressively. Because its foliage is unpalatable to livestock, the plant has been able to dominate some overgrazed pasture lands in the northern Rocky Mountains.

RELATED SPECIES: Rough cinquefoil (*Potentilla norvegica* L.) is an annual or biennial species with stout, hairy stems up to 3 feet (0.9 m) tall. Leaves have 3 leaflets that are hairy and green on both surfaces. Small, yellow flowers—less than 0.5 inch (1.2 cm) wide—have 5 petals that are shorter than the green sepals below them. Despite its species name, rough cinquefoil is native to both Europe and North America.

Sulfur cinquefoil growth habit

Lower leaves of sulfur cinquefoil

Close-up of sulfur cinquefoil flower

Sulfur cinquefoil flowers and developing fruits

Rough cinquefoil foliage

Close-up of rough cinquefoil flower

Galium mollugo L.   Smooth Bedstraw

SYNONYMS: wild madder, hedge bedstraw, false baby's breath, white bedstraw, whorled bedstraw

LIFE FORM: herbaceous perennial; with stems up to 3 feet (0.9 m) long

PLACE OF ORIGIN: Eurasia

VEGETATIVE CHARACTERISTICS: Smooth bedstraw is a weakly upright or sprawling plant with square stems that are smooth to the touch. Narrow, lance-shaped leaves lack petioles and are 0.5–1.5 inches (1.2–3.7 cm) long; 6–8 leaves are arranged in distinct whorls or layers at each node on the stem.

FLOWERS AND FRUIT: Smooth bedstraw produces a conspicuous terminal inflorescence consisting of numerous tiny, white flowers with 4 petals in late spring and summer, giving the plant a wispy appearance. Following insect-pollination, smooth bedstraw produces 2-lobed fruits that, unlike those of catchweed bedstraw (*below*), lack hooks on their surface.

GERMINATION AND REGENERATION: Seeds germinate readily in a variety of habitats; established plants produce yellow rhizomes and stolons that allow it to spread extensively.

HABITAT PREFERENCES: Smooth bedstraw grows best in shady or sunny, moist, nutrient-rich soil, but it can also grow in dry sandy sites with low pHs. In the urban environment, it is common along pathway edges, worn-out lawns and meadows, ornamental planting beds, and the moist edges of wetlands.

CULTURAL SIGNIFICANCE: Smooth bedstraw has a variety of uses in traditional medicine, including the treatment of kidney stones. As befits a member of the Madder Family, a red dye can be produced from its yellow rhizomes. Over time, it can take over old hayfields with poor soil and low pH, displacing more desirable species.

ECOLOGICAL FUNCTIONS: Disturbance-adapted colonizer of bare ground; erosion control through its tenacious roots; food and habitat for animals.

RELATED SPECIES: Catchweed bedstraw or **cleavers** (*Galium aparine* L.) is an annual species that appears to be native to North America, Europe, and Asia. Like the above species, 6–8 leaves are arranged in distinct whorls; unlike smooth bedstraw, its square stems, leaves, and fruits are all covered with short, downward-pointing "hooks" that are rough to the touch. Its tenacious fruits, in particular, readily cling to clothing and are an effective mechanism for long-distance dispersal by furry animals. Catchweed bedstraw produces small, white flowers in early summer on small axillary clusters (*cymes*); its sticky, prostrate or climbing stems can reach up to 3 feet (0.9 m) long. Extracts of its leaves have long been used as a "spring tonic" and a diuretic to treat kidney and bladder problems; the seeds, when roasted, have been used as a coffee substitute, and the dried leaves and stems were once used for mattress stuffing.

Field of smooth bedstraw in Vermont

Smooth bedstraw foliage

Smooth bedstraw in bloom

Catchweed bedstraw foliage

Catchweed bedstraw fruits covered with hooks

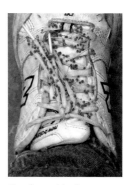

Catchweed bedstraw fruits stick to everything

Verbascum thapsus L.   Common Mullein

SYNONYMS: Aaron's rod, Jacob's staff, our lord's candle, flannel leaf, velvet dock, big taper, candlewick, great mullein, wooly mullein, lungwort, and so forth

LIFE FORM: **biennial**; up to 6 feet (1.8 m) tall

PLACE OF ORIGIN: Eurasia

VEGETATIVE CHARACTERISTICS: During its first year, mullein forms a flat rosette of woolly, gray-green leaves that stay green through the winter; in its second year, the plant produces a tall, unbranched flower stalk. Alternate leaves are oblong or lance-shaped; those at the base of the plant can be up to a foot (30 cm) long, those at the top are much smaller. Mullein produces a thick, fleshy taproot.

FLOWERS AND FRUIT: Sulfur-yellow flowers are produced from June through August on an unbranched terminal spike up to 18 inches (45 cm) long; vigorous plants sometimes produce branched flower stalks. Individual flowers have 5 fused petals, are about an inch in diameter, and are pressed tightly against the stalk. They are either insect- or self-pollinated. Fruit capsules are arranged spirally around the flower stalk, and each one contains several hundred tiny, dust-like seeds. A robust plant can produce more than 100,000 seeds.

GERMINATION AND REGENERATION: Mullein reproduces readily from seeds, which germinate in fall or spring, typically on freshly disturbed ground. They are dispersed only a short distance as the stalk moves back and forth in the wind. Mullein is the quintessential colonizer: It germinates from buried seeds following soil disturbance, lays down a large crop of seeds at the end of its second year, and then disappears until the next round of disturbance brings the seeds to the surface. In Europe, seeds buried for up to 500 years can still germinate.

HABITAT PREFERENCES: Mullein grows best on dry, sandy or gravelly soil that has been recently disturbed, especially on steep slopes where little else is growing. In the urban environment, it is common in vacant lots, rubble dumps, rock outcrops and stone walls, unmowed highway banks, and railroad rights-of-way.

CULTURAL SIGNIFICANCE: Dioscorides included mullein in his first-century herbal, *De Materia Medica*, and it has a long history in European folk medicine as a tea to treat chest colds, asthma, bronchitis, and kidney infections, and as a poultice to reduce swelling. Early settlers inserted the thick, fuzzy leaves into their shoes and clothing for extra insulation in the winter. Mullein was introduced into North America early on, probably for medicinal purposes. Darlington and Thurber (1859, 224) noted: "There is no surer evidence of a slovenly, negligent farmer, than to see his fields over-run with Mulleins. When neglected, the soil soon becomes so full of seeds that the young plants will be found springing up, in great number, for a long succession of years."

RELATED SPECIES: Moth mullein (*Verbascum blattaria* L.), another biennial with long-lived seeds from Eurasia, produces a rosette of non-hairy leaves with toothed margins in its first year and a tall flower stalk in its second. Flowers are an inch (2.5 cm) wide and either white or yellow with a reddish-purple center.

Common growing mullein on a bed of blacktop

Common mullein rosette at the end of its first year

Common mullein stalks at the end of their second year

Common mullein in flower

Common mullein seedlings on rubble soil

Close-up of common mullein flowers

Datura stramonium L. Jimson Weed

SYNONYMS: thorn apple, Jamestown weed, locoweed, stinkweed, devil's trumpet

LIFE FORM: annual; up to 5 feet (1.5 m) tall

PLACE OF ORIGIN: northern Mexico and southwestern United States

VEGETATIVE CHARACTERISTICS: Jimson weed typically forms a tall, well-branched plant with smooth, hollow stems that can be green or purple. Repeated forking of the primary stem is induced by the production of terminal flowers. Smooth, alternate leaves are large—up to 8 inches (20 cm) long—and are coarsely toothed along their margins; they emit a strong, unpleasant odor when crushed.

FLOWERS AND FRUIT: Jimson weed produces fragrant, funnel-shaped flowers, 3–4 inches (7.5–10 cm) long, which are white to pale violet in color; sepals are fused into a sheath that covers the lower half of the petals. Jimson weed begins producing flowers in June or July and continues doing so until it is killed by frost. Flowers are closed during the day and open at dusk and are commonly pollinated by hawkmoths. These are followed by 1–2 inch (2.5–5 cm) long, erect, egg-shaped pods covered with spines of various lengths; they are initially green, then brown, and split open when mature to release hard, black seeds.

GERMINATION AND REGENERATION: Seeds germinate with warm weather in late spring and, under favorable conditions, can form dense stands. Seeds can also remain dormant in the soil for many years and germinate when brought to the surface. Jimson weed often appears after agricultural soil containing its seeds is moved from one location to another for a landscape job. Plants are killed by the first hard frost, but erect, leafless stems persist long after death, identifiable by the remains of spiny seedpods.

HABITAT PREFERENCES: Jimson weed grows in a variety of soils in full sun but prefers ones that are alkaline and rich in nitrogen—such as those typically found in the vicinity of fenced livestock. In urban environments, jimson weed grows in a variety of disturbed, sunny habitats including walkway edges and riverbanks.

CULTURAL SIGNIFICANCE: All parts of jimson weed are considered toxic, and the plant has a long history of use for both its medicinal as well as psychotropic effects. It produces a number of powerful alkaloids, including atropine and scopolamine, which was once used to induce "twilight sleep" during childbirth and is still used to treat motion sickness. Various parts of the plant have traditionally been used to relieve the symptoms of asthma, rheumatism, and a host of other problems. Seeds, which are more toxic than other parts of the plant, have been used by numerous indigenous societies to induce visions. The plant was introduced from North America into Europe at an early date and can now be found in temperate and high-elevation subtropical zones throughout the world.

RELATED SPECIES: **Downy thorn apple** or **angel's trumpet** (*Datura innoxia* Mill.) is another annual species from Mexico that resembles jimson weed and produces the same toxic compounds. It has larger, white flowers up to 8 inches (20 cm) long and leaves covered with fine hairs. Rounded seed capsules are downward hanging rather than upright and are covered with spines of uniform length. The plant is commonly cultivated as an ornamental for its large, showy flowers, and seedlings can spread into adjacent landscapes.

ACEOUS DICOTS: **Solanaceae** (Nightshade Family)

Jimson weed brought in with fill soil in Connecticut

Dead stems and seedpods of jimson weed in winter

Jimson weed flowers and foliage

Jimson weed seedpod

Jimson weed flowers are approximately 4 inches long

Jimson weed flower head-on

SYNONYMS: West Indian nightshade, black nightshade

LIFE FORM: **summer annual** or **short-lived perennial**; up to 3 feet (90 cm) tall

PLACE OF ORIGIN: North America

VEGETATIVE CHARACTERISTICS: Eastern black nightshade is a low-branched plant with green stems that are smooth or slightly hairy, but not sticky to the touch. Alternate leaves are oval to egg-shaped and have irregular, blunt teeth along their margins and a dull purplish underside. Leaves are 1–5 inches (2.5–12.5 cm) long by half as wide and are typically covered with numerous tiny holes created by insects that feed on them.

FLOWERS AND FRUIT: White, star-shaped flowers, which are produced in late summer, are approximately 0.25 inch (6 mm) wide with yellow anthers; they are arranged in small, drooping clusters at intervals along the stem and can be either insect- or self-pollinated. Fruits are round, glossy black berries containing numerous seeds. Flowers and fruits form a true *umbel* with *pedicels* joined at a single point.

GERMINATION AND REGENERATION: Eastern black nightshade fruits are consumed by birds, and the seeds germinate readily beneath their roosting areas as well as in a variety of other habitats. Seeds buried in soil can remain viable for many years.

HABITAT PREFERENCES: Eastern black nightshade grows in a wide variety of disturbed habitats but prefers sunny, moist soil. In the urban environment, it is common in neglected ornamental landscapes, the margins of minimally maintained parks, vacant lots, rubble dumps, small pavement openings, and the compacted soil along walkways, stone walls, and rock outcrops.

CULTURAL SIGNIFICANCE: The foliage and green fruits of eastern black nightshade are poisonous to both livestock and people due to the presence of the alkaloid *solanine*. Fully mature, shiny black fruits are nontoxic and have been used to make pies and preserves. Although there are few references to the medicinal uses of eastern black nightshade, its close relative, European black nightshade (*below*), has an extensive history of use. Dioscorides included the plant in his first-century herbal, and it has a long history of use in Ayurvedic medicine in India as well as in Europe and Africa to treat inflammations, skin disorders, and eye problems. While Josselyn lists European black nightshade in *New-England's Rarities Discovered* (1672), he was probably observing the native American species with which it is often confused (Uva et al. 1997; Heiser 2003).

RELATED SPECIES: **European black nightshade (*Solanum nigrum* L.)** is native to Eurasia and is relatively common in eastern and central North America. It grows up to 18 inches (45 cm) tall and its fruits, which are dull black rather than shiny, are carried on a short, branched *raceme* rather than an *umbel*. The fact remains, however, that eastern and European black nightshades are difficult to tell apart, and taxonomists often lump 3 or 4 species and their hybrids together in the "black nightshade complex." Varieties of this species with edible fruits and foliage have been developed in subtropical regions of Africa and Asia.

Eastern black nightshade growth habit

Purplish leaf undersides of Eastern black nightshade

Eastern black nightshade flowers

Eastern black nightshade foliage is often damaged by insects

Eastern black nightshade fruits

Urtica dioica L. *subsp. gracilis* (Aiton) Seland. American Stinging Nettle

SYNONYMS: *Urtica procera, U. gracilis,* slender nettle, tall nettle

LIFE FORM: **herbaceous perennial**; up to 6 feet (1.8 m) tall

PLACE OF ORIGIN: northern and central North America

VEGETATIVE CHARACTERISTICS: This erect, unbranched plant has square, grooved, bristly stems. Narrow, opposite, egg- to lance-shaped leaves are 2–6 inches (5–15 cm) long with a rounded base and coarsely serrated margins. Leaves are arranged in pairs that are perpendicular to one another. Their lower surface is covered by silica-tipped, hollow, stinging hairs that release a skin irritant (formic acid) when touched. Compared to the European species (*below*), relatively few stinging hairs are found on its stems.

FLOWERS AND FRUIT: American stinging nettle produces pendulous chains of greenish-yellow (or sometimes pinkish) flowers in the axils of the uppermost leaves from late spring through summer. Individual flowers are small, lack petals, and are wind-pollinated; male and female flowers are separate from each other on the same plant (*monoecious*).

GERMINATION AND REGENERATION: Small seeds produced by the female flowers are dispersed passively as the plant blows in the wind; they germinate readily in moist soil. Established plants produce deeply rooted rhizomes that give rise to numerous new shoots and can form large colonies over time. Stinging nettles can be difficult to eliminate because of the tenacity of its rhizomes.

HABITAT PREFERENCES: Stinging nettles grows best in moist, nutrient-rich soil in either sun or shade. In the urban environment, it is common along the margins of freshwater wetlands, ponds, and streams; in moist soil at the edge of woodlands; and in roadside drainage ditches.

CULTURAL SIGNIFICANCE: Brushing against stinging nettle leaves results in a painful, burning sting that is not soon forgotten. When cooked, however, stinging nettles lose their sting and are safe to eat. In spring, young shoots can be collected (with gloves on) and boiled to make tea or, when pureed, a nutritious soup. The stem fibers can be used to make paper and a durable canvas-like cloth; the roots and leaves produce strong dyes.

RELATED SPECIES: **European stinging nettle (*Urtica dioica* L.)** is a smaller, weaker plant than its American cousin, and its broader leaves have a heart-shaped base. European stinging nettle is dioecious rather than monoecious, and the stinging hairs are located on both surfaces of the leaf blade. Dioscorides included stinging nettle in his first-century herbal, *De Materia Medica*, and indeed, every part of the plant seems to have some medicinal use. In the Middle Ages, people deliberately applied nettles to their skin not only as a form of religious penance (*urtication*) but also as a treatment for rheumatism and the loss of muscle strength. Tea made from the leaves is considered a tonic. Josselyn listed stinging nettle in *New-England's Rarities Discovered* (1672) under: "Of such plants as have sprung up since the English planted and kept cattle in New England," but he was probably referencing the North American rather than the European subspecies.

American stinging nettle
growth habit

Young American
stinging nettle shoots

Stand of American stinging nettles in spring

American stinging nettle flowers and foliage

European stinging nettle in flower

Verbena urticifolia L.   White Vervain

SYNONYM: nettle-leaved vervain

LIFE FORM: **annual, biennial, or short-lived perennial**; 3–5 feet (0.9–1.5 m) tall

PLACE OF ORIGIN: eastern North America

VEGETATIVE CHARACTERISTICS: White vervain produces square, light green stems densely covered with hairs. It has opposite, ovate to lanceolate leaves up to 6 inches (15 cm) long by 2.5 inches (5.7 cm) across with 2 inch (5 cm) long petioles; they are coarsely serrated along their margins and superficially resemble nettle leaves (hence its Latin name). Under good growing conditions, the plant is profusely branched and appears shrub-like.

FLOWERS AND FRUIT: White vervain's central stem terminates in a large, loose panicle of flowers up to 2 feet (60 cm) long and 1 foot (30 cm) across. This panicle is further subdivided into smaller 6 inch (15 cm) long branches that give the whole floral structure a loose, airy appearance. The white flowers are tiny, approximately 0.12 inch (3 mm) across, with 5 distinct petals, and are pollinated by a variety of bees. Only a few flowers open at one time, and the bloom period extends from midsummer into early fall; each flower can produce 4 tiny seeds. In late summer, white vervain presents a disorganized appearance with flowering spikes shooting out in different directions.

GERMINATION AND REGENERATION: Seeds can germinate in either spring or fall.

HABITAT PREFERENCES: White vervain typically grows in disturbed, semi-shaded areas with moist, rich soil, such as thickets, woodland and roadside edges, and beneath taller herbaceous plants; it occurs sporadically in the urban environment, but typically does not form large colonies.

CULTURAL SIGNIFICANCE: Native American tribes used the bitter root of white vervain to treat a variety of problems including profuse menstrual bleeding; early settlers used it to treat poison ivy blisters and hives.

RELATED SPECIES: **Blue vervain** or **swamp verbena** (***Verbena hastata*** L.) is an herbaceous perennial native to eastern North America that grows up to 5 feet (1.5 m) tall. It prefers moist, sunny fields, marshes, and meadows with rich soil and spreads by rhizomes. It produces hairy, square stems topped with a beautiful, candelabra-like inflorescence consisting of numerous slender spikes with small, blue-purple flowers from July through September; they are insect-pollinated and open sequentially from the bottom to the top of the spike. Doubly serrated, lance-shaped leaves are up to 6 inches (15 cm) long by an inch (2.5 cm) wide. Although not common in urban areas, blue vervain can be found in sunny, wet sites as well as those with disturbed, heavy soil.

White vervain in flower

The serrated leaf of white vervain

Close-up of white vervain flower

White vervain developing fruits

Blue vervain at the end of August in Connecticut

Mature seed heads of blue vervain in October

Viola sororia Willd.   Common Blue Violet

SYNONYMS: woolly blue violet, meadow violet, hooded blue violet, dooryard violet

LIFE FORM: **herbaceous perennial**; up to 8 inches (20 cm) tall

PLACE OF ORIGIN: eastern North America

VEGETATIVE CHARACTERISTICS: Common blue violet is a low-growing plant with glossy, heart-shaped leaves—up to 4 inches (10 cm) wide—arising from a basal crown. Leaf blades are smooth, highly variable in size, and have small, rounded teeth along their margins; the petioles are roughly twice as long as the leaf blades.

FLOWERS AND FRUIT: This species is well known for the flowers it produces from April through June; they come in a variety of shades of purple, blue, violet, and white—blue with a white throat being the most common. Flowers are approximately an inch (2.5 cm) wide, have 5 petals, and are carried on stalks that are about the same height as the leaves. In addition to the conspicuous, insect-pollinated flowers it produces in spring, common blue violet produces inconspicuous white flowers at or just below ground level later in the season. These hidden (*cleistogamous*) flowers never open and produce far more seeds than the aerial flowers.

GERMINATION AND REGENERATION: Common blue violet reproduces readily from seeds; established plants produce short rhizomes that give rise to new plants in the spring. Under favorable conditions, they can dominate large patches of lawn.

HABITAT PREFERENCES: Common blue violet grows best in cool, moist, shady soil, but it can also tolerate sites in full sun with dry soil. In the urban environment, it is common in minimally maintained lawns (it is tolerant of mowing), damp woodland thickets, roadside drainage ditches, and small holes in pavement.

CULTURAL SIGNIFICANCE: Native Americans used an infusion of common blue violet to treat a variety of minor aliments. Young leaves are edible in spring, as are the flowers, which can be eaten raw or candied.

RELATED SPECIES: **Jonny jump-up, wild pansy, heart's ease, or love-in-idleness (*Viola tricolor* L.)** is a European species that produces flowers with purple, yellow, and white coloration. It is often cultivated in gardens and occasionally escapes into the surrounding flower beds; it never strays too far from human habitation. In its native habitat, the plant is common along the disturbed edges of country roads and its flowers display a wide range of color patterns. This species is one of the progenitors of the cold-tolerant pansies sold in garden centers in early spring and was the inspiration for the notorious love potion in Shakespeare's *A Midsummer Night's Dream*.

Common
blue violet
growth habit

Underground flowers of common
blue violet

Common blue violet foliage

Jonny jump-ups in a neglected
flower garden

White-flowered form of
common blue violet

Typical common blue
violet flower

Allium vineale L.   Wild Garlic

Synonyms: field garlic, scallions, wild onion, crow garlic

Life Form: herbaceous perennial; up to 1–3 feet (30–90 cm) tall

Place of Origin: Europe, western Asia, and North Africa

Vegetative Characteristics: Wild garlic produces slender, hollow, grass-like leaves that look like chives and smell strongly of onions or garlic when crushed. The leaves, which are up to 8 inches (20 cm) long, arise from a small basal bulb and are circular in cross section. Wild garlic is a cool-season plant that grows vigorously in spring and early summer and dies back in July only to reappear in the fall. It is among the first plants to emerge in early spring.

Flowers and Fruit: Wild garlic produces round clusters of greenish or purplish flowers and tiny bulblets at the top of sturdy green stalks in May or June. The aerial bulblets, which typically far outnumber the flowers, grow to the size of a rice grain and sprout long, spidery leaves while still on the upright stalk. Wild garlic rarely—if ever—produces seeds in the Northeast.

Germination and Regeneration: As the bulblets expand, the flower stalk often bends over and touches the ground, thereby allowing them to take root and establish an independent existence. The mother plant regenerates in the spring from a perennial cluster of underground bulbs that are covered with a papery sheath.

Habitat Preferences: Wild garlic grows best in moist, nutrient-rich soils in sun or shade, but it can tolerate a wide range of habitats. It is common in minimally maintained public parks; disturbed woodlands; wet meadows; and the margins of freshwater wetlands, ponds, and streams.

Cultural Significance: John Bartram (1759, 452) described several reasons for not liking wild garlic: "Crow garlic—this is greatly loved by ye horses cows & sheep & very wholesome early pasture for them yet our people generally hates it because it makes ye milk butter cheese & indeed the flesh of those cattle that feeds much upon it taste so strong that we can hardly eat of it but for horses & young cattle it doth very well but our millers cant abide it amongst corn it clogs up their miss so that it is impossible to make good flower." Bulbs can be eaten in the spring.

Wild garlic in
early spring

Wild garlic in its typical urban niche

Wild garlic in the understory of an
Ailanthus grove

Wild garlic foliage grows back in the fall

Wild garlic bulblets sprouting
on top of the flower stalk

Commelina communis L.   Asiatic Dayflower

SYNONYM: common dayflower

LIFE FORM: **summer annual**; up to 2 feet (60 cm) tall by 2.5 feet (75 cm) wide

PLACE OF ORIGIN: eastern Asia

VEGETATIVE CHARACTERISTICS: The smooth, green stems of Asiatic dayflower have distinctly swollen nodes or joints; their growth is mainly horizontal but with upright shoot tips. Shiny, somewhat succulent leaves are lance-shaped to oblong and 2–5 inches (5–12.5 cm) long by 0.5–1.5 inches (1.2–3.7 cm) wide; they have parallel veins, and the sheaths at the base of the leaf blade clasp the stem. The root system is mostly adventitious and shallow-growing. The whole plant has a sprawling growth habit and dies with the first frost.

FLOWERS AND FRUIT: Flowers emerge from a folded, leafy bract; they are 0.5–1 inch (1.2–2.5 cm) wide and have 2 prominent blue petals on top and 1 smaller white petal below. The bisexual, insect-pollinated flowers are open for only a day—hence the name—and are produced throughout the summer and into the fall. The 3 cross-shaped, yellow structures at the center of the flower are sterile anthers, which make the flowers more attractive to potential pollinators.

GERMINATION AND REGENERATION: Seeds germinate readily; the stems root at the nodes where they touch the ground, allowing the plant to cover large areas.

HABITAT PREFERENCES: Asiatic dayflower grows best in moist, shady areas but, because of its succulent nature, also tolerates shady, dry conditions. It is common in minimally maintained public parks and gardens, the margins of freshwater wetlands and ponds, the rain shadow of buildings, and small pavement openings.

CULTURAL SIGNIFICANCE: Research from China has shown that Asiatic dayflower growing on toxic mine spoils bio-accumulates copper and other heavy metals—an observation that suggests it may be useful for phytoremediation purposes.

RELATED SPECIES: **Virginia spiderwort (*Tradescantia virginiana* L.)** is a native herbaceous perennial that can reach up to 2 feet (60 cm) tall. It grows best in moist soil in either sun or shade and readily escapes from landscape plantings. Its grass-like leaves are up to 12 inches (30 cm) long by an inch (2.5 cm) wide, clasp the stem, and are prominently grooved along the midvein. From late spring into summer, Virginia spiderwort produces clusters of blue to violet flowers at the tops of the stems subtended by 2 leaf-like bracts up to 6 inches (15 cm) long. Insect-pollinated flowers—which stay open for only a day—are approximately 1.5 inches (3.7 cm) across and roughly triangular in shape with 3 broad petals and 6 yellow anthers. Anther filaments are covered with fine, bluish hairs which, because of their large, clear cells, have been used in cytological and radiation studies. Virginia spiderwort hybridizes with **Ohio spiderwort (*T. ohiensis* Raf.)** to produce plants that are intermediate in the hairiness of their flower pedicels. In the days before herbicides, both species and the hybrid could often be found growing in ballast along sunny railroad tracks (Anderson 1952).

Asiatic dayflower flower

Asiatic dayflower set off by colored mulch

Asiatic dayflower growth habit

Asiatic dayflower foliage

Asiatic dayflower as an accidental ornamental

Virginia spiderwort in flower

Cyperus esculentus L.   Yellow Nutsedge

SYNONYMS: yellow nutgrass, nutsedge, chufa, galingale, umbrella sedge, rushnut, tigernut, nut flatsedge

LIFE FORM: herbaceous perennial; up to 2 feet (60 cm) tall

PLACE OF ORIGIN: southern Europe, Africa, India, and North America

VEGETATIVE CHARACTERISTICS: Shiny, grass-like leaves of yellow nutsedge have a distinctive yellowish-green tint that makes them stand out in the landscape. Leaf blades can be up to 18 inches (45 cm) long and 0.33 inch (8 mm) across; they are smooth with a conspicuous channel along the central vein. Upright, flowering stems are unbranched and triangular (all botany students learn that "sedges have edges").

FLOWERS AND FRUIT: Yellow nutsedge produces wind-pollinated flowers in August and September. The inflorescence consists of numerous clusters of shiny, yellowish-brown spikelets, 0.5–1.25 inches (1.2–3 cm) long, that are arranged horizontally and look like miniature bottle-brushes. They are produced near the top of the flower stalk and are subtended by thin, leafy bracts.

GERMINATION AND REGENERATION: Seeds germinate on moist, bare soil; established plants send out long, scaly rhizomes that terminate with succulent tubers ("chufas"). These grow outward the following spring and are responsible for the plant's ability to spread rapidly.

HABITAT PREFERENCES: Yellow nutsedge grows in a variety of disturbed habitats but prefers sunny, moist sites and nutrient-rich soil. It is common in landscape planting beds; poorly drained turf areas; and on the margins of freshwater wetlands, ponds, and streams. As a heat-loving species, it utilizes C_4 photosynthesis; it dies to the ground with frost but regrows in the spring from tubers.

CULTURAL SIGNIFICANCE: Underground tubers can be eaten raw, cooked, or dried and have a history of use dating back to the ancient Egyptians. Cultivated varieties are widely grown in Africa, India, and the Middle East for both human and livestock food, and in Spain and Central America to make a sweet, milk-like beverage called *horchata*. Yellow nutsedge is considered a problematic weed in many countries.

RELATED SPECIES: False nutsedge (*Cyperus strigosus* L.) is also native to North America and resembles yellow nutsedge except that its foliage lacks the yellowish cast, and it does not reproduce from tubers. It utilizes C_4 photosynthesis; prefers moist, sunny sites; and its seeds are capable of germinating after long periods of burial in the soil. Established plants expand slowly through the production of short, thick rhizomes.

Yellow nutsedge growth habit

Yellow nutsedge reproduces prolifically from rhizomes

Yellow nutsedge in the median strip

Yellow nutsedge infestation in a lawn

Yellow nutsedge inflorescences

Iris pseudacorus L.   Yellow Flag Iris

SYNONYMS: water flag, yellow iris, pale yellow iris, Jacob's sword, daggers, fleur-de-lis, flower-de-luce

LIFE FORM: **herbaceous perennial**; up to 5 feet (1.5 m) tall

PLACE OF ORIGIN: Europe, western Asia, and North Africa

VEGETATIVE CHARACTERISTICS: Yellow flag iris is a robust, clump-forming marsh plant with stiff, erect, sword-shaped leaves that are approximately 3 feet (0.9 m) long, an inch (2.5 cm) wide, with a distinctive bluish-green cast. As the season advances, the leaves are beaten down by the weather and become prostrate.

FLOWERS AND FRUIT: Clusters of showy, bright yellow to cream-colored flowers, 2.5–3 inches (6.7–7.5 cm) wide, are produced on tall, branched stalks in June. Following pollination by bees, cylindrical green fruits develop; they can be up to 3 inches (7.5 cm) long and contain 3 stacks of flattened, D-shaped seeds.

GERMINATION AND REGENERATION: Seeds germinate readily in sunny, moist soils; established plants spread vigorously from thick, branching rhizomes that can form large, dense mats of foliage.

HABITAT PREFERENCES: Yellow flag iris grows well in a wide variety of freshwater and brackish wetlands, including the margins of ponds and streams; standing water up to 9 inches (23 cm) deep; and poorly drained, acidic soils. It grows well in either full sun or light shade and can tolerate periods of drought.

ECOLOGICAL FUNCTIONS: Stream and riverbank stabilization; nutrient absorption in wetlands; food and habitat for wildlife.

CULTURAL SIGNIFICANCE: The flower is the inspiration of the fleur-de-lis, the heraldic emblem of the kings of France. Yellow flag iris also has traditional medicinal uses. Dioscorides included it in his first-century herbal, *De Materia Medica*, and in traditional European medicine the rhizome was used as a powerful cathartic as well as for a number of other less dramatic purposes, including as a cosmetic to remove bruises. Because of its high tannic acid content, the root was also used for tanning leather. In recent times, yellow flag iris has been planted in water treatment wetlands for its capacity to absorb nutrients and heavy metals (phytoremediation). The plant was introduced into North America in the mid-1800s, probably for ornamental purposes, and has spread spontaneously into native wetlands. Many states list it as an invasive species.

Yellow flag iris in bloom in June

Yellow flag iris growth habit

Yellow flag iris emerging in early spring

Yellow flag iris in late fall

Yellow flag iris flower

Yellow flag iris fruits

Juncus tenuis Willd.   Path Rush

SYNONYMS: *Juncus macer,* slender rush, wire grass, poverty rush

LIFE FORM: **evergreen perennial**; up to 1 foot (30 cm) tall

PLACE OF ORIGIN: North and Central America

VEGETATIVE CHARACTERISTICS: Path rush is a clump-forming, grass-like plant with tough, round, hollow, dark green stems. The scraggly basal leaves, which are about half the height of the flowering stems, are flat with inwardly rolled edges and persist through the winter. The stems, which are extremely flexible and spring back quickly after being stepped on or driven over, often have a diagonal orientation. The shallow, fibrous root system is remarkably tenacious.

FLOWERS AND FRUIT: Path rush produces clusters of inconspicuous greenish-brown flowers near the top of the stiff stalks in late spring and early summer. Individual flowers, which have 3 petals and 3 long sepals, are subtended by long, leafy bracts. Following wind-pollination, small fruit capsules develop and split into 3 parts at maturity, releasing hundreds of orange-brown seeds.

GERMINATION AND REGENERATION: Path rush seeds become sticky when wet, are transported by animals and people (as well as tire treads), and germinate in sunny sites with poor soil; established plants increase in size by means of short rhizomes. Seeds are capable of germinating after long periods of burial in the soil.

HABITAT PREFERENCES: As the common name suggests, this species thrives in compacted, drought-prone soil and is common along heavily traveled walkways as well as in pavement cracks with pedestrian or vehicular traffic. It also grows in heavy, wet soils and in dirt roads paved with gravel.

CULTURAL SIGNIFICANCE: Path rush is one of the few plants in this book for which humans have found no uses. It seems to have been inadvertently introduced into Europe in the late 1700s and has become well established on that continent, primarily in compacted soils along roadways.

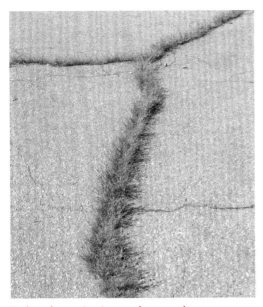

Path rush growing in a roadway crack

Path rush growth habit

Path rush in flower

Path rush in the fall

Developing fruits of path rush

Bromus tectorum L.   Downy Brome

SYNONYMS: cheatgrass, downy chess, early chess, thatch grass, gas station grass, drooping brome

LIFE FORM: **winter annual**; up to 2 feet (60 cm) tall

PLACE OF ORIGIN: Eurasia and North Africa

VEGETATIVE CHARACTERISTICS: All parts of downy brome are covered with soft, silky hairs that give it a "fuzzy" appearance. This cool-season grass begins producing leaves in early spring; they are twisted as they emerge, making it look like they are spiraling upward. Flat, hairy leaf blades can be up to 8 inches (20 cm) long. The fibrous root system is extremely efficient at absorbing soil moisture, especially where very little is available.

FLOWERS AND FRUIT: Downy brome produces a distinctive drooping inflorescence from early spring through midsummer depending on the time of seed germination; the spikelets, which consist of 3–8 florets with prominent awns, are 0.5–1 inch (1.2–2.5 cm) long. Following wind-pollination, flower heads develop into soft, shiny, purplish seed heads that turn tawny brown at maturity. Seed spikelets have long bristles that facilitate their dispersal by animals (including people).

GERMINATION AND REGENERATION: Seeds germinate in the fall or spring, depending on local conditions; seeds that germinate in the fall are able to grow roots as long as the soil temperatures are above freezing.

HABITAT PREFERENCES: This drought-tolerant species flourishes in sandy or compacted soils throughout the urban environment. It is common in pavement cracks and openings, highway banks and median strips, and along railroad tracks.

CULTURAL SIGNIFICANCE: The species name, *tectorum*, means "of roofs" in Latin and indicates that this plant was traditionally used in Europe for thatching. Downy brome was first reported in North America in 1861 in New York. It moved west during the 1890s—both as a plant cultivated for livestock forage and as a weed following railroad tracks. By the 1930s, downy brome had become a major problem in the Great Basin region, helped along by drought and overgrazing. It is now considered a major invasive species in western rangelands where it displaces native grasses and promotes fire.

RELATED SPECIES: **Smooth** or **Hungarian brome (*Bromus intermis* Leyss.)** is a cool-season perennial grass that was introduced in the 1880s as a forage plant for livestock and escaped to become a common roadside species. It gained popularity during the 1930s because of its drought tolerance. Smooth brome grows up to 4 feet (1.2 m) tall and produces a nodding, purple-tinged inflorescence in late spring or early summer. At maturity, the seed heads droop conspicuously from the weight of the large, single-seeded fruits.

Downy brome in its typical urban niche

Downy brome in the city

Downy brome colonizing a pile of asphalt

Developing seeds of downy brome

Mature seed heads of downy brome

Smooth brome seed heads

Dactylis glomerata L.   Orchardgrass

SYNONYMS: cock's foot, cat grass

LIFE FORM: **evergreen perennial**; up to 4 feet (1.2 m) tall

PLACE OF ORIGIN: Europe

VEGETATIVE CHARACTERISTICS: Orchardgrass is a tall, clump-forming, cool-season plant that stays green through the winter and produces new leaves in early spring. Bluish-green leaf blades are up to 12 inches (30 cm) long and distinctly creased in the middle, both as they emerge from the bud and when fully expanded. The densely clustered stems arise from perennial crowns and can form large clumps, or tussocks. Roots are very tenacious and can be pulled up only when the soil is wet.

FLOWERS AND FRUIT: Orchardgrass produces wind-pollinated flowers from spring through midsummer in spikelets that are crowded in 1-sided clusters at the end of stiff, upright branches. The clumped, fluffy appearance of the flowers and the seed heads makes this grass easy to identify in both summer and fall.

GERMINATION AND REGENERATION: Seeds germinate readily on sunny, bare soil and can also germinate after being buried in the soil for many years. Established plants enlarge slowly through the growth of short rhizomes, often forming circular patches with a dead patch at their center.

HABITAT PREFERENCES: Orchardgrass is an adaptable species that is tolerant of drought, shade, and low soil fertility. It is extremely common in minimally maintained public parks, unmowed highway banks and median strips, vacant lots and abandoned property, and along sunny woodland edges.

ECOLOGICAL FUNCTIONS: Tolerant of roadway salt and compacted soil; erosion control on slopes; food and habitat for wildlife.

CULTURAL SIGNIFICANCE: Orchardgrass was originally introduced from Europe as a forage crop for domestic animals and is still a major component of many hayfields in the Northeast. Darlington and Thurber (1859, 379) explained the origin of its common name: "This grass also possesses the additional advantage of thriving well in the shade of trees, and answers a very good purpose in orchards." For whatever reason, cats seem to be fond of eating this grass.

Leaf collar of
orchardgrass

Orchardgrass
in bloom

Orchardgrass
dominates the
vast urban
meadows of
Detroit

Orchardgrass in early spring

Orchardgrass with mature
seeds

Dichanthelium clandestinum (L.) Gould   Deer-tongue Grass

SYNONYMS: *Panicum clandestinum*, riverside panic grass

LIFE FORM: **evergreen perennial**; up to 4 feet (1.2 m) tall

PLACE OF ORIGIN: eastern North America

VEGETATIVE CHARACTERISTICS: Upright to sprawling stems of this cool-season native grass have broad, stiff, lance-shaped leaves that are 3–4 inches (7.5–10 cm) long by an inch (2.5 cm) wide. The base of the leaf clasps the stem, and the sheath is distinctly hairy. Leaves turn brown in the fall and persist on the stems all winter, making this plant easy to identify. Because of its wide leaves, the plant looks like a miniature bamboo.

FLOWERS AND FRUIT: Deer-tongue grass produces flowering stalks from May through September. Wind-pollinated flowers produced early in the season usually emerge completely, whereas those produced later in the season typically fail to emerge fully from their surrounding sheaths and are self-pollinated (hence the species name, *clandestinum*).

GERMINATION AND REGENERATION: Seeds germinate in sun or shade and are capable of germinating after long periods of burial. Established plants expand in size through the growth of rhizomes and can form large clumps.

HABITAT PREFERENCES: Deer-tongue grass seems to grow best in moist, partly shaded conditions but is also found in dry, sunny locations. It is tolerant of acid soils and is common along roadside drainage ditches; in the shady margins of wetlands, streams, and ponds; and in the compacted soil of vacant lots. Deer-tongue grass is unusual in that it has broad leaves and utilizes C_3 photosynthesis.

ECOLOGICAL FUNCTIONS: Disturbance-adapted colonizer of bare ground; food and habitat for wildlife; erosion control on slopes and acidic mine tailings.

RELATED SPECIES: Japanese stiltgrass (*Microstegium vimineum* (Trin.) A. Camus) is an annual, warm-season grass (C_4 photosynthesis) from Japan whose seeds were inadvertently introduced into Tennessee in 1919 as packing material. It has spread southward and northward since then, arriving in southern New England in the 1980s. It grows up to 3 feet (0.9 m) tall and produces alternate, light green leaves that are 2–3 inches (5–7.5 cm) long and approximately 0.5 inch (1.2 cm) wide. The species flowers in September and matures its seeds about a month later. Remarkable for its ability to grow and reproduce in the shade, it can form dense stands in forest understories. It is also quite salt tolerant and is common along the edges of country roads. The spread of this plant is facilitated by the fact that it is not eaten by deer. Japanese stiltgrass is currently more common in rural and suburban areas than in cities, but this could change in the future.

Deer-tongue grass growing along a heavily salted road

Leaf collar of deer-tongue grass

Deer-tongue grass foliage

Deer-tongue grass in winter

The inflorescence of deer-tongue grass often does not fully emerge from the leaf axil

Japanese stiltgrass growing in a shady woodland

SYNONYMS: small crabgrass, fingergrass, twitchgrass

LIFE FORM: **summer annual**; up to 1 foot (30 cm) tall

PLACE OF ORIGIN: Europe

VEGETATIVE CHARACTERISTICS: Smooth crabgrass seedlings appear in late spring and early summer and grow rampantly until killed by frost. The plant has a prostrate or spreading growth habit and typically does not produce roots at the leaf nodes. Leaf blades and sheaths are relatively smooth (hairless), as are the stems, which often develop a reddish coloration when stressed. The root system of established plants is extremely tenacious.

FLOWERS AND FRUIT: Skinny, wind-pollinated flower heads are composed of numerous finger-like spikes alternately arranged at the top of a slender stem. The spikes are green when they emerge in late summer, then change to purple, and finally become tan as the seeds mature.

GERMINATION AND REGENERATION: Seeds germinate readily in late spring and early summer and can remain viable in the soil for several years.

HABITAT PREFERENCES: Smooth crabgrass is remarkably drought tolerant and grows robustly in the heat of summer. It is common in trampled lawns in minimally maintained public parks and residential landscapes, unmowed highway banks and median strips, small pavement openings, and sidewalk cracks. As a warm-season species, smooth crabgrass utilizes C_4 photosynthesis.

CULTURAL SIGNIFICANCE: Smooth crabgrass was once used as a forage grass in the South where cool-season grasses did not grow well. It typically dominates patchy lawns—especially those damaged by grubs or drought—where it is often the only grass that is still green in August.

RELATED SPECIES: **Large** or **hairy crabgrass (*Digitaria sanguinalis* (L.) Scop.)** is common in sunny, disturbed habitats, including lawns, ornamental planting beds, pavement cracks, and median strips. It is more robust and upright than smooth crabgrass, and its stems root at the nodes. The leaves and stems of this species are densely covered with fine hairs, and the inflorescence consists of 4–10 finger-like spikes clustered at the top of the stem. It is not as common in lawns as smooth crabgrass because it is taller and less tolerant of mowing. In Europe, the seeds of large crabgrass have been used for food. **Goosegrass (*Eleusine indica* (L.) Gaertn.)** is a robust summer annual, probably native to Africa, with C_4 photosynthesis. It resembles crabgrass but has flattened stems that radiate out from a central point and are silvery white at their base. If allowed to grow freely, goosegrass can become 2–3 feet (60–90 cm) tall, but when mowed it forms a ground-hugging mat. Its flowers are produced in midsummer in jagged, finger-like clusters, and its dry seed heads persist through the winter. Seeds are edible.

Smooth crabgrass in a typical urban niche

Smooth crabgrass on the street

Smooth crabgrass tolerates foot traffic

Smooth crabgrass at the end of the season

Goosegrass growth habit

Large crabgrass inflorescence

Echinochloa crus-galli (L.) Beauv. Barnyard grass

SYNONYMS: cockspur grass, billion dollar grass, Japanese millet, barn-grass

LIFE FORM: **summer annual**; up to 5 feet (1.5 m) tall

PLACE OF ORIGIN: tropical Asia

VEGETATIVE CHARACTERISTICS: Barnyard grass is a vigorous, upright grower that produces multiple stems with adventitious prop roots at their base. Smooth leaf blades have a distinct light green to white midrib and can be 4–8 inches (10–20 cm) long and 0.2–0.8 inch (5–20 mm) wide; the sheaths are smooth, flattened, and reddish-purple. The shallow root system makes the plant relatively easy to pull up.

FLOWERS AND FRUIT: Erect, terminal flower panicles are greenish-red, coarsely branched, and covered with bristles. They are wind-pollinated and produced throughout the summer, and the seed heads range from dark green to purple at maturity.

GERMINATION AND REGENERATION: Individual plants can produce tens of thousands of seeds per year; they germinate readily once the temperatures warm up in late spring.

HABITAT PREFERENCES: Barnyard grass grows best in nutrient-rich, moist soil with full sun but can also tolerate shady, dry conditions. It is extremely salt tolerant and is common along agricultural edges, disturbed wet areas, waste dumps, minimally maintained landscape plantings, roadsides, and moist drainage ditches. As a warm-season species, barnyard grass utilizes C_4 photosynthesis.

CULTURAL SIGNIFICANCE: Barnyard grass is a common and very serious weed in agricultural fields and disturbed areas in tropical Asia and Africa where it can significantly reduce crop yields. It also grows along the edges of lakes, streams, and rivers throughout much of North America and Europe. Seeds are readily consumed by wildlife and have been used as food by humans in times of famine. Darlington and Thurber (1859, 403) explained that the common name of this species derives from the fact that "it is apt to abound along the drains of crude liquid flowing from barn-yards—and in spots which are usually designated as 'wet and sour.'"

Barnyard grass growth habit

Leaf collar of barnyard grass

Barnyardgrass foliage

Barnyard grass mature seed head

Barnyard grass young seed head

Barnyard grass mature seed

Elymus repens (L.) Gould   Quackgrass

Synonyms: *Agropyron repens, Elytrigia repens,* couch grass, quick grass, scratch grass, twitch grass, quake grass, scotch grass, creeping wheat

Life Form: evergreen perennial; up to 4 feet (1.2 m) tall

Place of Origin: Eurasia and North America

Vegetative Characteristics: Quackgrass is a "running" grass with stems (*culms*) emerging at some distance from one another. Bluish-green leaves are rolled in the bud as they emerge and have sharp-pointed tips; leaf blades can be up to a foot (30 cm) long. Crushed foliage has the distinctive smell of fresh wheat. The plant is evergreen and can establish large colonies that are difficult to eradicate. The name "quack grass" is a corruption of "quick grass," which refers to the fact that it emerges in very early spring.

Flowers and Fruit: Quackgrass produces a 2–4 inch (5–20 cm) long inflorescence at the end of a tall stalk; wind-pollinated flowers (*spikelets*) are arranged alternately along the axis. The seed head is an erect spike.

Germination and Regeneration: Seeds germinate readily on open ground; established plants spread rapidly by means of sharp-pointed, white rhizomes that can produce both roots and shoots at the regularly spaced nodes.

Habitat Preferences: Quackgrass is common in minimally maintained landscape areas with rich soil and full sun. It is also found in abandoned grasslands and meadows, small pavement openings, roadsides and median strips, and along railroad tracks.

Ecological Functions: Disturbance-adapted colonizer of bare ground; tolerant of roadway salt and compacted soil; erosion control on slopes.

Cultural Significance: Quackgrass was commonly planted in the 19th century as a forage crop valued for its ability to green up quickly and to produce sweet hay. Over time, its aggressive, spreading habit became apparent and the plant was redefined as a weed that needed to be controlled. Rhizomes collected in the spring were used in European traditional medicine as a diuretic to treat a variety of urinary problems and to purify the blood, and they have also been used to make bread in times of famine.

Quackgrass growth habit

Quackgrass in bloom

Quackgrass growing along a sidewalk

Quackgrass in its urban niche

Quackgrass rhizomes

Quackgrass mature seed

Eragrostis pectinacea (Michx.) Nees.   Tufted Lovegrass

SYNONYM: Carolina lovegrass

LIFE FORM: summer annual; 2–24 inches (5–60 cm) tall

PLACE OF ORIGIN: North America, except Alaska

VEGETATIVE CHARACTERISTICS: In good soil, tufted lovegrass forms loose, open clumps up to 2 feet (60 cm) tall; in urban sidewalk cracks, the clumps are denser and usually less than half that height. This sprawling grass produces thin, smooth stems; narrow leaves; and thin, flat flower stalks. New stems arise from the base and often lie flat on the ground but do not form adventitious roots. When in flower, the whole plant has a shiny appearance and glistens in the sun.

FLOWERS AND FRUIT: Tufted lovegrass produces minute, wind-pollinated flowers on short spikelets that are arranged in a loose terminal inflorescence. They are produced from July through September, typically within a month or so after germination, followed by wispy seed heads.

GERMINATION AND REGENERATION: Seeds germinate readily in sunny, moist or dry locations in late spring or early summer, after the soil has warmed up.

HABITAT PREFERENCES: This inconspicuous, heat-tolerant grass is extremely common in the urban environment, especially in sidewalk and pavement cracks. It also grows in other disturbed sites in full sun, including vacant lots, rock outcrops, minimally maintained ornamental plantings, highway banks, and median strips. As a warm-season species, tufted lovegrass utilizes C_4 photosynthesis, which allows it to tolerate periods of hot, dry weather.

RELATED SPECIES: Purple lovegrass (*Eragrostis spectabilis* (Pursh) Steud.) is another native, warm-season (C_4 photosynthesis) species that is common along infrequently mowed highway edges with frequent salt applications. It grows to approximately 18 inches (45 cm) tall and, if not mowed, produces beautiful, dull red inflorescences that become extremely conspicuous in August and September.

Tufted lovegrass in a sidewalk crack

Tufted lovegrass growing only on the short sides of bricks

Tufted lovegrass in blacktop

Purple lovegrass dominating a highway median strip in fall

Purple lovegrass in seed

Festuca arundinacea Schreb.   Tall Fescue

SYNONYMS: *Lolium arundinacea, Festuca elatior*, coarse or meadow fescue, tall rye-grass, Kentucky fescue

LIFE FORM: **evergreen perennial**; up to 2 feet (60 cm) tall

PLACE OF ORIGIN: Europe

VEGETATIVE CHARACTERISTICS: Tall fescue is easily recognized in the landscape by its broad, dark green leaves, whose glossy upper surfaces glisten in bright sunshine. Leaves are much coarser than most other lawn grasses and remain conspicuously green through the winter. As they emerge in spring, the leaves are rolled (as opposed to folded) in the bud and lack a prominent midrib; on robust plants, the leaf blades can be up to 2 feet (60 cm) long by 0.5 inch (12 mm) wide. The fibrous root system is deep and extremely tenacious.

FLOWERS AND FRUIT: Tall fescue produces relatively thin, wind-pollinated flowers in spring and early summer on stalks that grow to approximately 2 feet (60 cm) tall; they persist in a dry state into the fall.

GERMINATION AND REGENERATION: Tall fescue germinates readily from seeds; established plants are long-lived and expand in size by producing new shoots from the base (*tillers*).

HABITAT PREFERENCES: This species is highly tolerant of drought and low-fertility soils and grows well in either sun or shade; it is common on steep grassy slopes, minimally maintained turf areas, and along the compacted margins of paths and walkways.

ECOLOGICAL FUNCTIONS: Tall fescue is tolerant of roadway salt and compacted soil and resists erosion on slopes. The leaves of this species contain a fungal endophyte, which increases its tolerance of environmental stress.

CULTURAL SIGNIFICANCE: Tall fescue was introduced from Europe in the late 1800s as a pasture grass. Because of its toughness, it has become the most commonly planted grass for football fields and golf courses—covering literally tens of millions of acres across North America. It is also widely planted for erosion control on steep, difficult sites and is a common ingredient in drought-tolerant lawn seed mixes.

RELATED SPECIES: **Red fescue (*Festuca rubra* L.)** is a clump-forming, cool-season grass with wiry leaves that are much smaller and finer than those of tall fescue. It is native to Europe and is now widely used in drought-tolerant lawn seed mixes. It naturalizes freely in semi-shaded, dry sites at the edge of emergent woodlands. Including its flower stalks, red fescue can grow to approximately 2 feet (60 cm) tall. The cultivar 'Chewings' is widely planted because of its rapid growth.

Tall fescue foliage

Tall fescue holding a bank in winter

Tall fescue in flower

Tall fescue planted for erosion control

Widely spaced clumps of red fescue

Red fescue in flower

Muhlenbergia schreberi J.F. Gmel.   Nimblewill

SYNONYMS: wiregrass, drop-seed

LIFE FORM: **herbaceous perennial**; up to 2 feet (60 cm) tall

PLACE OF ORIGIN: North America

VEGETATIVE CHARACTERISTICS: Nimblewill produces numerous slender, wiry stems that branch freely from the base and sprawl across the ground, eventually forming large clumps. Alternate gray-green leaves are narrow, sharp pointed, and only about 2 inches (5 cm) long. The prostrate stems typically form shallow-growing, adventitious roots at their lower nodes.

FLOWERS AND FRUIT: Nimblewill is one of the last grasses to bloom, usually in September and October. The slender inflorescences are 2–8 inches (5–20 cm) long and produced at the ends of both the terminal and the side branches with a somewhat wispy appearance, often with a slight purplish tint. Wind-pollinated flowers are inconspicuous.

GERMINATION AND REGENERATION: Seeds germinate readily; once the plant is established, the horizontal stems root at the nodes and can form large clumps.

HABITAT PREFERENCES: Nimblewill tolerates a wide range of moisture and light conditions from full sun to dense shade. In shady, dry sites (e.g., the rain shadow of buildings) it is quite spindly; on moist, rich soil it can be very robust. It tolerates compaction and is a common component of minimally managed lawns and parks where mowing encourages its spread. As a warm-season species, nimblewill utilizes C_4 photosynthesis.

RELATED SPECIES: **Bermuda grass** or **couch grass (*Cynodon dactylon* (L.) Pers.)** is a warm-season perennial that spreads vegetatively by long underground or aboveground runners that root at the jointed nodes. As a native of East Africa, it utilizes C_4 photosynthesis and is well adapted to warm climates, but cold-hardy varieties can be found growing spontaneously as far north as New England. Bermuda grass tolerates a wide range of ecological conditions, including drought, physical disturbance, and soils with either a high or a low pH or contaminated with salt. The 3–7 flower spikes are produced at the top of the stems from July through October; these skinny "fingers" superficially resemble those of crabgrass (*Digitaria* spp.), but they are shorter and thicker and radiate out from a central point at the top of the stem. Because of its adaptability, Bermuda grass is an important forage plant in parts of the southern United States. It is also commonly used for lawns in tropical and subtropical countries because, as the locals say, "The more you cut it, the more it grows." It is common in overgrazed portions of the shortgrass plains of the Serengeti in Africa.

Nimblewill in flower in late summer

Nimblewill in spring

Nimblewill along a sidewalk

Nimblewill growth habit

Bermuda grass in a New Orleans roadway

Bermuda grass in flower

Panicum dichotomiflorum Michx.   Fall Panicum

Synonyms: smooth witchgrass, fall panicgrass

Life Form: summer annual; from 6 inches (30 cm) to 4 feet (1.2 m) tall

Place of Origin: eastern North America

Vegetative Characteristics: The growth habit of fall panicum can vary from totally prostrate (when growing in poor soil) to upright (when growing in rich soil); clumps produce multiple stems with a distinctive zigzag form. Leaf blades are mostly smooth with a conspicuous white midrib; the leaf sheaths are also smooth and typically are reddish-purple. The fibrous root system is extremely tenacious. A hard frost kills fall panicum, leaving behind a mass of tawny foliage.

Flowers and Fruit: Fall panicum produces panicles of wind-pollinated flowers from late summer through fall, followed by loose, spreading seed heads that turn purple and then brown with the first frost.

Germination and Regeneration: Seeds germinate readily on bare ground in late spring.

Habitat Preferences: While fall panicum grows best in sunny sites with moist, loamy soil, it is also quite drought tolerant and is common in a variety of disturbed urban habitats with compacted or clay soil, including pavement cracks and underneath highway guardrails where blacktop and concrete meet. As a warm-season grass, fall panicum utilizes C_4 photosynthesis.

Ecological Functions: Disturbance-adapted colonizer of bare ground; food and habitat for wildlife.

Cultural Significance: The presence of sprawling fall panicum plants in parking lot and sidewalk cracks creates the impression of neglect.

Related Species: Switchgrass (*Panicum virgatum* L.) is a large, clump-forming perennial grass native to North America. It often grows in dry soils along sandy roadsides and at the upland edge of salt marshes and other wetlands. Its dead, leafy stems stand upright through the winter. It is a warm-season species that utilizes the C_4 photosynthesis. Because of its ability to produce abundant biomass—it can grow up to 7 feet (2.1 m) tall—switchgrass is being promoted for cultivation on marginal land as a source of cellulosic ethanol. The plant has become a popular ornamental in recent years, and many select cultivars are available in the nursery trade.

Fall panicum growth habit

Fall panicum in bloom

Fall panicum in an abandoned parking lot

Fall panicum in its urban niche

Switchgrass growing in a median strip

Switchgrass after a hard frost

Phalaris arundinacea L.   Reed Canarygrass

SYNONYMS: lady grass, gardener's garters

LIFE FORM: **herbaceous perennial**; up to 5 feet (1.5 m) tall

PLACE OF ORIGIN: Eurasia and western North America; most of the plants growing in eastern North America originated in Europe

VEGETATIVE CHARACTERISTICS: Stout, upright stems of reed canarygrass are bluish-green, and the smooth leaves appear to be arranged mostly on one side of the stem. Blades are broad and flat, 4–10 inches (10–25 cm) long by 0.5–1.5 inches (1.2–3.7 cm) wide, and have a pointed tip and rough edges. Leaves are rolled as they emerge from the bud. It is a cool-season species that begins growing in early spring.

FLOWERS AND FRUIT: Tall flower stalks extend well above the leaves from mid-May through August. The inflorescence itself is approximately 6 inches (15 cm) long and initially compressed into a tight spike; this expands into a loose panicle of wind-pollinated flowers that mature into a straw-colored seed head.

GERMINATION AND REGENERATION: Seeds germinate readily; established plants spread by means of thick underground rhizomes that produce dense clusters of stems.

HABITAT PREFERENCES: Reed canarygrass grows best in moist, sunny sites and can become dominant in degraded wetlands and along the margins of freshwater marshes, streams, and ponds. It can also be found in shady, moist woods, in roadside drainage ditches, and dry urban sites. Under conditions to its liking, it can form dense stands.

ECOLOGICAL FUNCTIONS: Nutrient absorption in wetlands; tolerant of roadway salt and compacted soil; food and habitat for wildlife; erosion control on stream banks.

CULTURAL SIGNIFICANCE: Although many states list reed canarygrass as an invasive species, it is still cultivated in the Northeast as a forage crop for cattle. The old-fashioned variegated variety known as "ribbon grass" (the cultivar 'Picta') is widely grown as an ornamental. Research from England suggests that the species has phytoremediation potential on brownfield sites.

Reed canarygrass in bloom

Reed canarygrass
rhizomes

Leaf collars of reed
canarygrass

Reed canarygrass growth habit

Reed canarygrass (*foreground*)
in a degraded wetland with
yellow flag iris (*midground*) and
common reed (*background*)

Phleum pratense L.   Timothy

SYNONYMS: herd's grass, cat's tail

LIFE FORM: **herbaceous perennial**; 3–4 feet (0.9–1.2 m) tall

PLACE OF ORIGIN: Europe

VEGETATIVE CHARACTERISTICS: Stems of timothy are light green and smooth, and have a distinct, bulb-like thickening at the base (*corm*) from whence its fibrous root system emerges. Smooth, gray-green leaves have flat blades—3–9 inches (7.5–22.5 cm) long—that taper to a sharp point and are widely separated from one another on the stem. New leaves, as they emerge in early spring, are rolled in the bud as opposed to being folded.

FLOWERS AND FRUIT: In June and July, timothy produces a compact, cylindrical spike of flowers 2–4 inches (5–10 cm) long and less than 0.5 inch (1.2 cm) wide at the end of a tall, leafless stalk. Inflorescences are purplish in color and remarkably soft to the touch during the pollination period; they mature into light brown seed heads. Individual shoots die after they set seed.

GERMINATION AND REGENERATION: Seeds germinate readily in the fall, shortly after dispersal. Established plants increase in size through the production of new shoots from the swollen base; the entire clump typically survives 4–5 years.

HABITAT PREFERENCES: Timothy is a cool-season grass that grows best in moist, nutrient-rich soil in full sun. In the urban landscape, it typically has a scattered distribution in minimally maintained grasslands and meadows; along roadsides; and on the margins of freshwater wetlands, ponds, and streams. It is a common component of erosion control seed mixes.

CULTURAL SIGNIFICANCE: Timothy is important as forage and as a pasture grass for making hay. Between 1870 and 1910, before the internal combustion engine was invented, it was a major cash crop because it provided nutritious fuel for the horses that propelled machines and carriages. The first record of timothy growing in the United States dates from 1711 when Jonathan Herd found it growing along the Piscataqua River in New Hampshire, most likely as an accidental introduction from England. Its use for hay was first publicly promoted in 1720 by a farmer named Timothy Hanson; hence the common name. The naturalist John Bartram was another early proponent of planting timothy for feeding livestock. In areas where timothy is abundant, it can be a source of allergy-inducing pollen in early summer.

Timothy inflorescences

Timothy shedding its pollen

Bulb at base of timothy stalk

Bluish-green timothy foliage

Mature timothy seed head

Timothy in bloom

SYNONYMS: *Phragmites communis*, *P. maximus*, *P. phragmites*, reedgrass, giant reed, marsh reed

LIFE FORM: **herbaceous perennial**; up to 15 feet (4.5 m) tall

PLACE OF ORIGIN: Asia, Europe, and North America

VEGETATIVE CHARACTERISTICS: This tall, long-lived grass produces round, hollow stems, and alternate leaves are conspicuously arranged in a single plane. Smooth, blue-green leaf blades are 1–2 feet (30–60 cm) long by approximately an inch (2.5 cm) wide with rough margins and prominent veins.

FLOWERS AND FRUIT: The inflorescence is a conspicuous plume-like panicle, 6–15 inches (15–42 cm) long, that is purple at first and then, following wind-pollination, changes to tawny brown as the seeds mature. The dry stalks with their tan leaves and fluffy seed heads remain standing through winter and can become a fire hazard during periods of spring drought.

GERMINATION AND REGENERATION: Seeds are dispersed by wind and can germinate under a variety of moisture and light conditions. More commonly, common reed spreads by rhizomes and can form huge populations of genetically uniform stems.

HABITAT PREFERENCES: Common reed is a disturbance-adapted plant with C_3 photosynthesis that can dominate sunny wetlands—both fresh and salt water—throughout the temperate world. It is common in moist soils wherever wetland drainage has been impeded by road construction, in vacant city lots and rubble dumps, at the margins of freshwater wetlands and streams, and in brackish tidal marshes.

ECOLOGICAL FUNCTIONS: Nutrient absorption (mainly nitrogen and phosphorus) in degraded wetlands; tolerant of roadway salt; food and habitat for wildlife (muskrats); stream, river, and coastal shoreline stabilization.

CULTURAL SIGNIFICANCE: In various European and Middle Eastern countries, the dry stalks of common reed were traditionally used as building material—mainly for roof thatching—and the young shoots were fed to domestic animals. In Russia, the plant has been used as a source of cellulose for making paper. Native Americans used many parts of the plant for food: the boiled rhizomes in winter, the young shoots as a vegetable in spring, and the seeds as a grain in fall. In the Southwest, the stems were used in the construction of adobe huts and as shafts for arrows. Europeans used the stems to make quills for pens. Common reed is widely planted throughout the temperate world for soil stabilization and, in constructed wetlands, for the tertiary purification of wastewater from sewage treatment plants (phytoremediation).

RELATED SPECIES: *Phragmites australis* **subspecies *americanus*** is native to the East Coast but is relatively rare and not nearly as salt tolerant as the aggressively spreading variety—**subspecies *australis***—that was unintentionally introduced from Europe in the early 1800s (Saltonstall 2002). Indeed, the European variety of *Phragmites* is one of the most widely distributed plants in the world and is listed as an invasive species by many states.

Common reed in the New Jersey Meadowlands

Common reed is remarkably drought tolerant

New shoots of common reed coming up through old ones

Phragmites australis variety *gigantissima* growing up to 20 feet (6 m) tall in Boston's Back Bay Fens

Common reed in flower

Poa annua L.   Annual Bluegrass

SYNONYMS: spear grass, dwarf meadow grass, winter grass, six-weeks grass, cause-way grass, walk-grass

LIFE FORM: **annual**; up to 1 foot (30 cm) tall

PLACE OF ORIGIN: Europe

VEGETATIVE CHARACTERISTICS: This small, tufted grass has smooth stems that root wherever they touch the ground. Leaf blades are up to 2 inches (5 cm) long, smooth, and light yellowish-green with flattened sheaths; the new leaves are folded in bud. Being a cool-season grass, the whole plant dies in the midsummer heat. Members of the genus *Poa* can be distinguished from other grasses by their prow-shaped leaf tips.

FLOWERS AND FRUIT: In early spring and again in the fall, annual bluegrass produces short, dense panicles of silvery white, wind-pollinated flowers 1–3 inches (2.5–7 cm) long. It is an opportunist species that can mature its seeds within a few days of pollination (hence the name "six-weeks grass").

GERMINATION AND REGENERATION: Seeds germinate in late summer, early fall, or spring, depending on the location and the weather. Seeds buried in the soil can survive for up to 30 years.

HABITAT PREFERENCES: Annual bluegrass is common in compacted lawns and walkways, minimally maintained public parks, vacant lots, rubble dumps, small pavement openings and cracks, and unmowed highway banks and median strips.

ECOLOGICAL FUNCTIONS: Disturbance-adapted colonizer of bare ground; remarkably tolerant of roadway salt, soil compaction, and foot traffic.

CULTURAL SIGNIFICANCE: Annual bluegrass is one of the most common grasses in temperate urban environments, and it is one of the few non-native, vascular plants that grow in the maritime areas of Antarctica.

RELATED SPECIES: **Canada bluegrass (*Poa compressa* L.)** is a spring-flowering perennial species from Europe with a clump-forming growth habit and fine, narrow, blue-green foliage. It typically grows 1–2 feet (30–60 cm) tall with wiry stems that are distinctly flattened rather than round. Stems are topped with short, compact flower panicles that are 2–3 inches (5–7.5 cm) long. Canada bluegrass is common on dry, infertile sites, and it typically goes dormant in summer (C_3 photosynthesis).

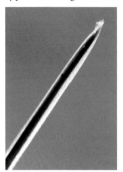

Prow-shaped leaf tip typical of the genus *Poa*

Annual bluegrass in its urban niche

Annual bluegrass approximately 2 inches (5 cm) tall

Annual bluegrass growing in a cobblestone gutter

Canada bluegrass stems and foliage

Canada bluegrass in flower

Poa pratensis L.   Kentucky Bluegrass

Synonyms: smooth meadow grass, Junegrass

Life Form: semi-evergreen perennial; up to 2 feet (60 cm) tall

Place of Origin: Europe and North America (Canada)

Vegetative Characteristics: Kentucky bluegrass produces tufts or clumps of smooth leaves that are folded in the bud, and blades are approximately 4 inches (10 cm) long. The tips of the blades are shaped like the prow of a boat, a feature common to all members of the genus *Poa*.

Flowers and Fruit: Kentucky bluegrass produces open panicles of green, wind-pollinated flowers in May or June (hence the name "Junegrass"); flower heads occupy the top 20% or so of the flower stalk, which typically is approximately 2 feet (60 cm) tall. Stalks are round or oval in cross section with distinct, blackish "joints" at the upper, leafless nodes. The seed heads turn tan and then brown as they mature. In some varieties, the flowers develop without fertilization (*apomixis*).

Germination and Regeneration: Seeds germinate in spring or fall in moist, sunny or shady locations; established plants produce creeping rhizomes that allow the clumps to expand quickly. Plants typically go dormant in summer when exposed to warm temperatures or drought, or both. Once established, Kentucky bluegrass can persist in minimally maintained turf areas for many years.

Habitat Preferences: Given its reputation for high maintenance in lawns—lots of water and good soil—it is surprising to find Kentucky bluegrass growing spontaneously in a variety of disturbed sites, including sunny woodland edges and urban meadows. It prefers limestone soils and is tolerant of road salt.

Cultural Significance: Kentucky bluegrass has a long history of cultivation in pastures and is the most widely grown lawn grass in the Northeast. Its popularity stems from its deep green color and its capacity to spread quickly and produce thick clumps. Although it is apparently native to parts of western North America, Kentucky bluegrass in eastern North America was introduced from Europe, probably around 1685 by William Penn. The musical genre "bluegrass" was named for this species, which dominates the pastures of rural Kentucky.

Related Species: Wood bluegrass or **wood meadowgrass (*Poa nemoralis* L.)** is an evergreen European species that grows to approximately 2 feet (60 cm) tall and is typically found in the shady understory and along the edges of woodlands. Because it does not produce rhizomes or stolons, wood bluegrass has a more clumped growth habit than Kentucky bluegrass has and is intolerant of mowing. Its densely clustered flower stalks are produced in June and arch over at a 45° angle as they mature, producing a graceful effect in the summer landscape.

Unmowed Kentucky bluegrass lawn

Kentucky bluegrass in flower

Clumped growth habit of wood bluegrass in early spring

Wood bluegrass growth habit

Wood bluegrass in winter

Wood bluegrass in flower

Setaria viridis (L.) Beauv.   Green Foxtail

SYNONYM: green brittle grass

LIFE FORM: **summer annual**; from 4–36 inches (10–90 cm) tall

PLACE OF ORIGIN: Eurasia

VEGETATIVE CHARACTERISTICS: Green foxtail produces vigorous new shoots, or tillers, at the base and forms a dense cluster of stems by the end of the growing season. Light green leaves are rolled in the bud, and the blades can be up to 10 inches (25 cm) long.

FLOWERS AND FRUIT: Light green, wind-pollinated flowers are produced in mid-to-late summer. They are approximately 3 inches (7.5 cm) long and are arranged in tight, upright spikes at the end of tall, leafless stalks. Individual flowers are green and are subtended by numerous stiff bristles. The whole plant dies in the fall, leaving the tawny brown seed heads standing tall above a skirt of withered leaves.

GERMINATION AND REGENERATION: Seeds are dispersed when the mature heads shatter; they germinate in late spring or early summer when the soil starts to warm up. Seeds can germinate after long periods of burial in the soil.

HABITAT PREFERENCES: Green foxtail is quite common in the urban environment, especially in compacted soil and full sun. It grows in small pavement cracks and openings, minimally maintained public parks and landscape planting beds, and unmowed highway banks and median strips. As a warm-season grass, green foxtail utilizes C_4 photosynthesis.

CULTURAL SIGNIFICANCE: Green foxtail is the wild progenitor of foxtail millet (*Setaria italica*), which the Chinese domesticated some 8,000 years ago. Green foxtail has many uses in Asian and European traditional medicine and was intentionally introduced into the United States as a forage grass in the early 1800s.

RELATED SPECIES: **Yellow foxtail (*Setaria pumila* (Poir.) Roem. & Schult.)** produces numerous tillers and forms clumps that are approximately 3 feet (0.9 m) tall. Its upright seed heads, up to 4 inches (10 cm) long, are distinctly yellow and bristly. It is common along roadways and abandoned grasslands and stands out in August when its flower heads create a yellow haze on the landscape. **Giant or Chinese foxtail (*Setaria farberi* Herrm.)** was introduced inadvertently in the 1930s from Asia. It is a larger, more robust plant than the other two foxtails, reaching up to 6 feet (1.8 m) tall, with a curving rather than an erect seed head that can be up to 7 inches (17.5 cm) long. Both yellow and giant foxtails grow best in moist, rich soil—agricultural fields—and are more common in the countryside than in the city. Both of these species are warm-season annuals that utilize C_4 photosynthesis.

Green foxtail in seed

Green foxtail at the end of its life

Giant foxtail growth habit

Giant foxtail at the end of its life

Giant foxtail in seed

Yellow foxtail in seed

Smilax rotundifolia L.   Roundleaf Greenbrier

SYNONYMS: catbrier, bullbrier, hellfetter, blasphemy vine, horsebrier

LIFE FORM: **perennial woody liana (vine)**; with stems up to 20 feet (6 m) long

PLACE OF ORIGIN: eastern North America

VEGETATIVE CHARACTERISTICS: Roundleaf greenbrier is a low-climbing or sprawling vine whose slender green stems are studded with stiff, sharp thorns that make the plant extremely unpleasant to encounter in the woods—almost like running into barbed wire. Glossy leaves are alternate, simple, rounded to heart-shaped, and roughly 2–6 inches (5–15 cm) long and wide; they have smooth margins and 3–5 distinctly parallel veins. The plant climbs by means of the tendrils that terminate the leaf stipules. Plants typically shed their leaves in late fall; the leafless stems maintain their olive-green coloration through the winter.

FLOWERS AND FRUIT: Roundleaf greenbrier is a dioecious species with separate male and female individuals. Greenish flowers—about 0.25 inch (6 mm) across—are produced in small axillary clusters from late spring through early summer and are pollinated by a variety of insects; bluish-black fruits approximately 0.25 inch (6 mm) in diameter develop in September.

GERMINATION AND REGENERATION: Fruits are dispersed by birds, and the seeds germinate in a variety of habitats. Following disturbance, established plants produce new stems from deep-growing underground rhizomes, leading to the formation of dense, impenetrable thickets.

HABITAT PREFERENCES: Roundleaf greenbrier grows best on moist, sunny sites but also tolerates dry, shady conditions. It is common along roadsides, at the edges of thickets, in the understory of low, moist woods, and on chain-link fences that border woodlands.

ECOLOGICAL FUNCTIONS: Food and habitat for wildlife; tolerant of roadway salt and compacted soil; stream and riverbank stabilization.

CULTURAL SIGNIFICANCE: Young shoots and leaves are edible in spring, and the roots produce a gelatin that can be used to thicken soups.

Roundleaf greenbriar flowers

Thicket of roundleaf greenbriar stems in winter

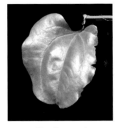

Roundleaf greenbriar leaf

Roundleaf greenbriar on a chain-link fence

Roundleaf greenbriar tendrils

Roundleaf greenbriar fruits

Typha latifolia L.   Common Cattail

SYNONYMS: cat-o-nine-tails, cattail flag, blackamoor, candlewick, water torch, bulrush, punks, corndog grass

LIFE FORM: **herbaceous perennial**; 6–9 feet (1.8–2.7 m) tall

PLACE OF ORIGIN: North America and Eurasia

VEGETATIVE CHARACTERISTICS: Common cattail is an erect, grass-like perennial that grows from 4–8 feet (1.5–3 m) tall with flat, sword-shaped leaves arising from the base to create a fan-like appearance. Bluish or gray-green blades can reach up to 7 feet (2 m) long by an inch or so (2.5 cm) wide. New shoots and roots arise in early spring from a stout, creeping rhizome.

FLOWERS AND FRUIT: Common cattail produces brownish flowers in dense, cylindrical spikes from May through July. Two distinct types of flowers occur on the same stalk, separated by less than 0.5 inch (1.2 cm): males—located above—shed their yellow pollen to the wind and quickly disappear; and females—located below—develop over the course of the summer into the hotdog-shaped "cat tails" that release their fluffy, wind-dispersed seeds over the course of the fall and winter. A single, tightly packed cattail holds approximately 200,000 seeds!

GERMINATION AND REGENERATION: Seeds germinate in moist soil; established plants are strongly rhizomatous and form large stands.

HABITAT PREFERENCES: Common cattail grows best along the sunny margins of brackish marshes, freshwater wetlands, and ponds. It is also common in drainage ditches along the margins of highways.

ECOLOGICAL FUNCTIONS: Excess nutrient absorption in wetlands; tolerant of roadway salt; food and habitat for wildlife; stream and riverbank stabilization.

CULTURAL SIGNIFICANCE: The peeled rhizomes and tender young shoots of common cattail can be eaten raw in salads or cooked as a vegetable. Native Americans used all parts of the plant to treat a variety of skin injuries and ailments as well as diarrhea and dysentery. The plant also has a number of industrial uses based on its processed fibers.

RELATED SPECIES: Narrowleaf cattail (*Typha angustifolia* L.) can grow to approximately 6 feet (2 m) tall and produces sword-shaped leaves that are narrower than those of the common cattail—only about 0.5 inch (1.2 cm) wide. On the flowering stalks, the gap between male flowers (top) and female flowers (below) is 2–4 inches (5–10 cm) as opposed to only 0.5 inch (1.2 cm) for common cattail. In general, narrowleaf cattail grows in deeper water than does common cattail and is more tolerant of salt, fertilizer runoff, and alkaline soils. The range of this native species has increased dramatically over the past century.

Common cattails dispersing their seeds

A common cattail marsh in Boston

Common cattail foliage

Common cattail and purple loosestrife

Common cattail flowers in spring, with males located above female

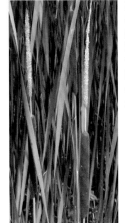

Narrowleaf cattail flowers in spring, with males located above females

APPENDIX 1. Urban Habitats and Their Preadapted Plants

Below are descriptions of nine urban environmental conditions or habitats that urban plants are preadapted to, followed by examples of the species that flourish in these circumstances.

1. **Drought-tolerant, tap-rooted annuals with a prostrate growth habit** are adapted to growing in sunny, sandy soils as well as limestone or granite cliffs and rock outcrops. In cities, these plants are preadapted to growing in sidewalk cracks with compacted soil and to being trampled by people. Examples include carpetweed (*Mollugo verticillata*), purslane (*Portulaca oleracea*), spotted spurge (*Euphorbia maculata*), and prostrate knotweed (*Polygonum aviculare*). A number of fibrous-rooted, trampling-tolerant grasses and rushes also inhabit sidewalk cracks, including smooth crabgrass (*Digitaria ischaemum*), tufted lovegrass (*Eragrostis pectinacea*), and path rush (*Juncus tenuis*).

2. Weathering associated with **the widespread use of de-icing salts** along walkways and roadways produces high pH microhabitats along their edges that are typically colonized by species that come from grasslands or floodplains with limestone-rich soils (calciphiles). Examples include mugwort (*Artemesia vulgaris*), spotted knapweed (*Centaurea stoebe*), chicory (*Cichorium intybus*), and dandelion (*Taraxacum officinale*). Along highways where salt applications are heavy and maintenance is minimal, woody species such as tree-of-heaven (*Ailanthus altissima*) and black locust (*Robinia pseudoacacia*) dominate unmowed slopes. **Nitrogen deposition associated with acid precipitation** (see below) also contributes to the deterioration of concrete and creates localized pockets of alkalinity.

3. **Atmospheric deposition of nitrogen** in urban areas is roughly twice that of rural areas—on the order of 8.7 kg per hectare per year. Deposition rates are highest in the spring and closely correlated with vehicle emissions (Decina et al. 2017). Atmospheric nitrogen deposition can promote robust growth in nitrophile species, including summer annuals such as lambsquarters (*Chenopodium album*), common mallow (*Malva neglecta*), and amaranths (*Amaranthus* species).

4. **Early-successional trees** that typically grow on eroded slopes, disturbed forest edges, or river floodplains typically show a strong capacity for vegetative regeneration from basal sprouts, rhizomes, or root suckers (Del Tredici 2001). In urban environments, species that can sprout vigorously from the base or from roots are selectively favored by the activities of maintenance crews who constantly hack plants back. Examples include boxelder (*Acer negundo*), silver maple (*Acer*

saccharinum), tree-of-heaven (*Ailanthus altissima*), northern catalpa (*Catalpa speciosa*), white mulberry (*Morus alba*), black locust (*Robinia pseudoacacia*), and various elms (*Ulmus* species).

5. River corridors and floodplains provide an abundance of **high-light and high-disturbance edge habitats**. They are characterized by fluctuating water levels and heavy soils, and the trees that grow naturally under such conditions are preadapted to grow on urban streets where soil is often compacted and low in oxygen. Examples of native bottom-land species that perform well as street trees include silver maple (*Acer saccharinum*), river birch (*Betula nigra*), green ash (*Fraxinus pensylvanica*), honey locust (*Gleditsia triacanthos*), sweetgum (*Liquidambar styraciflua*), pin oak (*Quercus palustris*), and the once ubiquitous American elm (*Ulmus americana*).

6. The term **urban heat island** describes the fact that urban areas trap and store more heat during the daytime than do adjacent non-urban areas with less paving and fewer buildings. This stored heat is released at night, typically keeping cities significantly warmer in both summer and winter than the surrounding countryside. Examples of trees that respond positively to warmer (and drier) urban microclimates include tree-of-heaven (*Ailanthus altissima*), silk tree (*Albizia julibrissin*), and northern catalpa (*Catalpa speciosa*). Among herbaceous plants, those that utilize the C_4 **photosynthesis pathway**—which are more productive and water-use efficient in hot, dry environments than C_3 plants—are significantly more abundant and diverse in urban areas than in non-urban areas (Duffy and Chown 2016). Examples of urban C_4 plants include purslane (*Portulaca oleracea*), spotted spurge (*Euphorbia maculata*), carpetweed (*Mollugo veticiallata*), various pigweeds (*Amaranthus* species), and the so-called warm-season grasses in the genera *Cynodon, Digitaria, Eragrostis, Panicum,* and *Setaria*.

7. **Disturbance-adapted winter annuals** that germinate in fall and flower in spring are common in the urban landscape. They are disproportionately of European origin and are increasing in abundance under the combined influences of climate change and urbanization. Mild fall weather extends for a longer period of time in cities than it does in the surrounding countryside, which allows the rosette stage of the plant to become larger. In addition, warmer, shorter winters allow the rosettes to start growing earlier in spring, which promotes increased flower and seed production. Examples of urban winter annuals include garlic mustard (*Alliaria petiolata*), yellow rocket (*Barbarea vulgaris*), shepherd's purse (*Capsella bursa-pastoris*), and hairy bittercress (*Cardamine hirsuta*), all members of the Mustard Family (Brassicaceae).

8. In urban environments, **chain-link fences** provide unique opportunities for the establishment and growth of vines. Not only do they provide climbing plants with the opportunity to overtop the competition, they also provide protection from the ravages of maintenance crews. Woody vines (lianas) that grow well on chain-link fences include porcelain berry (*Ampelopsis glandulosa* var. *brevipedunculata*), oriental bittersweet (*Celastrus orbiculatus*), and Japanese honeysuckle (*Lonicera japonica*); herbaceous perennial vines include hedge bindweed (*Calystegia sepium*) and black swallowwort (*Vintoxicum nigrum*). Chain-link fences also act as safe sites

for the establishment and survival of tree saplings by protecting them from mowers and weed whackers.

9. Opportunistic woody plants that leaf out as soon as temperatures warm up in the spring or that hold their leaves until a hard frost kills them in late fall are able to take advantage of the **increased growing season length** that is a result of both climate change and the urban heat island effect. Many of the Asian and European shrubs growing in the understories of eastern North American forests leaf out earlier and retain their leaves longer than do the native species that grow in the same area (Fridley 2013). Examples of early leaf-out species include Japanese barberry (*Berberis thunbergii*), honeysuckles (*Lonicera* species), wild apples (*Malus* species), and multiflora rose (*Rosa multiflora*); late leaf-retaining species include glossy buckthorn (*Frangula alnus*), common buckthorn (*Rhamnus cathartica*), and black alder (*Alnus glutinosa*).

APPENDIX 2. Plants Treated in This Book That
Are Included in Dioscorides' *De Materia Medica*

The Greek physician Pedanius Dioscorides (AD 40–90) practiced medicine in Rome under the emperor Nero and wrote the five-volume *De Materia Medica*, a compendium of medicinal plants, animals, and minerals that remained in active use into the 1600s. Modern research has demonstrated that many of the plants Dioscorides mentioned contain medically active compounds and were effective herbal remedies in their day. Among the plants mentioned in the earliest known edition of this book, from the 5th century (Gunther 1959), are the following.

Achillea millefolium	Yarrow
Arctium minus	Burdock
Calystegia sepium	Hedge bindweed
Capsella bursa-pastoris	Shepherd's purse
Chelidonium majus	Greater celandine
Daucus carota	Queen Anne's lace
Glechoma hederacea	Ground ivy
Hypericum perforatum	Common St. Johnswort
Iris pseudacorus	Yellow flag iris
Leonurus cardiaca	Motherwort
Medicago sativa	Alfalfa
Persicaria maculosa	Ladysthumb
Populus alba	White poplar
Portulaca oleracea	Purslane
Rhamnus cathartica	Common buckthorn
Saponaria officinalis	Bouncing Bet
Solanum nigrum	Black nightshade
Sonchus oleraceus	Annual sowthistle
Tussilago farfara	Coltsfoot
Urtica dioica subsp. *dioica*	European stinging nettle
Verbascum thapsus	Common mullein
Vintoxicum rossicum	Pale swallowwort

APPENDIX 3. European Plants Listed by Josselyn as Growing Spontaneously in New England in the Seventeenth Century

Plants listed under the category "Of such plants as have sprung up since the English planted and kept cattle in New England."

Capsella bursa-pastoris	Shepherd's purse
Malva neglecta	Common mallow
*Plantago major**	Broadleaf plantain
Polygonum aviculare	Prostrate knotweed
Rumex crispus	Curly dock
Senecio vulgaris	Groundsel
*Solanum nigrum**	Black nightshade
Sonchus oleraceus	Annual sowthistle
Stellaria media	Chickweed
Taraxicum officinale	Dandelion
*Urtica dioica**	Stinging nettle

*These species could well have been North American species that Josselyn assumed came from Europe.

Josselyn mistakenly listed the following plants as native to both England and New England.

Arctium minus	Common burdock
Chelidonium majus	Greater celandine
Chenopodium album	Lambsquarters
Glechoma hederacea	Ground ivy
Hypericum perforatum	Common St. Johnswort
Linaria vulgaris	Yellow toadflax
Persicaria maculosa	Ladysthumb
Portulaca oleracea	Purslane
Tanacetum vulgare	Common tansy
Veronica arvensis	Corn speedwell

Source: From *New England's Rarities Discovered* (1672) by John Josselyn with modern identifications provided by Tuckerman (1865).

APPENDIX 4. Species Suitable for a Cosmopolitan Urban Meadow

Below is a list of herbaceous perennials described in the book, arranged by plant family, that are suitable for growing as part of a cosmopolitan urban meadow. They are tough, adaptable plants with ornamental attributes. They are not meant to be planted together; rather, they provide a palate of species that can be mixed and matched with other herbaceous species—both native (n) and non-native—to create an attractive meadow on typical urban soil.

Species	Common name	Family
Asclepias syriaca (n)	Milkweed	Apocynaceae
Achillea millefolium	Yarrow	Asteraceae
Cichorium intybus	Chicory	Asteraceae
Leucanthemum vulgare	Oxeye daisy	Asteraceae
Solidago juncea (n)	Early goldenrod	Asteraceae
Solidago sempervirens (n)	Seaside goldenrod	Asteraceae
Symphyotrichum pilosum (n)	White heath aster	Asteraceae
Tanacetum vulgare	Tansy	Asteraceae
Saponaria officinalis	Bouncing Bet	Caryophyllaceae
Tradescantia virginiana (n)	Virginia spiderwort	Commelinaceae
Lotus corniculatus	Birdsfoot trefoil	Fabaceae
Trifolium hybridum	Alsike clover	Fabaceae
Trifolium pretense	Red clover	Fabaceae
Dichanthelium clandestinum (n)	Deer-tongue grass	Poaceae
Eragrostis spectabilis (n)	Purple lovegrass	Poaceae
Festuca rubra	Red fescue	Poaceae
Verbena hastata (n)	Blue vervain	Verbenaceae

APPENDIX 5. Shade-Tolerance Ratings
of the 40 Trees Covered in This Book

Shade-intolerant species (55%)

Acer saccharinum	Silver maple
Ailanthus altissima	Tree-of-heaven
*Albizia julibrissin**	Silktree
Alnus glutinosa	Black alder
Betula nigra	River birch
Betula pendula	European silver birch
Betula populifolia	Gray birch
Catalpa speciosa	Northern catalpa
Elaeagnus angustifolia	Russian olive
Gleditsia triacanthos	Honey locust
Juglans nigra	Black walnut
Morus alba	White mulberry
*Paulownia tomentosa**	Princess tree
Populus alba	White poplar
Populus deltoides	Eastern cottonwood
Populus tremuloides	Quaking aspen
*Pyrus calleryana**	Callery pear
Quercus palustris	Pin oak
Rhus glabra	Smooth sumac
Rhus typhina	Staghorn sumac
Robinia pseudoacacia	Black locust
Salix x *sepulcralis*	Weeping willow

Intermediate shade-tolerant species (35%)

Acer negundo	Box elder
*Acer pseudoplatanus**	Sycamore maple
Acer rubrum	Red maple
Carya cordiformis	Bitternut hickory
Celtis occidentalis	Hackberry
Fraxinus americana	White ash
Fraxinus pennsylvanica	Green ash

Intermediate shade-tolerant species (35%) *continued*

Malus pumila	Common apple
Prunus serotina	Black cherry
Prunus virginiana	Choke cherry
Quercus rubra	Red oak
Ulmus americana	American elm
Ulmus pumila	Siberian elm
Zelkova serrata*	Zelkova

Very shade-tolerant species (10%)

Acer platanoides	Norway maple
Phellodendron amurense*	Amur corktree
Rhamnus cathartica	Common buckthorn
Taxus cuspidata*	Japanese yew

Source: Based on data in *Michigan Trees* by B. V. Barnes and W. H. Wagner Jr., University of Michigan Press, 2004.

*not listed in *Michigan Trees*.

APPENDIX 6. Key Characteristics of Important Plant Families

Asteraceae (Compositae): Sunflower Family
– Many individual flowers (*florets*) are combined on a common receptacle
  to form a composite inflorescence.
– Two basic types of flowers make up the composite heads: ray or ligulate florets
  with strap-shaped, petal-like appendages, and are often infertile; tube-shaped
  disk florets that lack petal-like appendages and are usually fertile.
– Some, such as the oxeye daisy, have both ray and disk florets.
– Some, such as dandelion, chicory, and prickly lettuce, have only ray florets
  with conspicuous "petals" and milky sap.
– Some, such as burdock, groundsel, and pineapple weed, have only disk florets
  without conspicuous "petals" and have clear sap.
– Flowers pollinated by insects or self-pollinated.

Brassicaceae (Cruciferae): Mustard Family
– Herbaceous plants, often biennial with basal rosettes.
– Simple, alternate leaves are often divided, cleft, or lobed.
– Flowers have 4 petals forming a cross (mainly yellow or white).
– Flowers pollinated by insects or self-pollinated.
– Seedpod is a 2-chambered capsule that is either long and narrow (*silique*)
  or more or less rounded (*silicle*). Seedpods often explode at maturity, leaving
  only a clear membrane (*placenta*) behind.
– Sap contains pungent oils.
– Root systems of most species lack mycorrhizae.

Caryophyllaceae: Pink Family
– Leaves are opposite, entire, and joined together at the base of the stem
  to form swollen nodes.
– Flowers have 5 petals, often notched at the tip, which makes it look like
  10 petals.
– Sepals are often united to form an inflated tube.
– Flowers pollinated by insects or self-pollinated.
– Fruits consist of a dry capsule filled with small seeds.
– Root systems of many species lack mycorrhizae.

Euphorbiaceae: Spurge Family

  – Flowers often have colored bracts that look like petals.
  – Flowers pollinated by wind or insects.
  – Many species exude latex-rich, milky sap when broken.
  – Leaves and branching are alternate.
  – Plants are often, but not always, toxic to humans.

Fabaceae (Leguminosae): Pea or Bean Family

  – Consists of three subfamilies: the Mimosadae, with "powder-puff" flowers; the Caesalpinoideae, with separate, more or less equal petals; and the Papalinoideae, with the typical pea-type flower consisting of 5 petals: 1 is upright (the banner), 2 are lateral (the wings), and 2 are fused and project outward (the keel).
  – All subfamilies produce a fruit that is a pea-like pod (a legume) containing one or more seeds and that splits open on two sides.
  – Alternate leaves are often pinnately compound.
  – Most members of the Papalinoideae have root nodules that fix nitrogen.

Lamiaceae (Labiatae): Mint Family

  – Stems are square and foliage is aromatic.
  – Opposite, simple leaves have toothed margins and a decussate arrangement.
  – Flowers have 5 fused petals that form 2 prominent lips: an upper one with 2 lobes and a lower one with 3 lobes.
  – Flowers typically pollinated by insects.

Poaceae (Graminae): Grass Family

  – Stems are round with distinct nodes and hollow internodes.
  – Leaves are alternate and often arranged in a single plane.
  – Leaves consist of a blade that projects out from the stem and a sheath that wraps around the stem; the blades have parallel venation.
  – Flowers are subtended by bracts and aggregated into clusters (*spikelets*).
  – Flower spikelets lack distinctive petals and sepals and are wind-pollinated.
  – New shoots arise from rhizomes or stolons.
  – Cool-season species have C_3 photosynthesis; warm-season species are C_4.

Polygonaceae: Smartweed Family

  – Stem is swollen at the leaf nodes, forming distinct "knots."
  – Leaves are alternate, simple, without teeth along their edges.
  – Membranaceous stipules (*ocrea*) encircle the stem at the nodes.
  – Flowers are small with reddish sepals and without petals.
  – Flowers pollinated by wind or by insects.
  – One-seeded fruits (*achenes*) usually have 3 distinct wings to facilitate wind or water dispersal.
  – Root systems of many species lack mycorrhizae.

Ranunculaceae: Buttercup Family

- Bisexual flowers with numerous parts independently attached below
  a superior ovary (i.e., the flower parts are not fused).
- No set number of floral parts; multiple pistils at the center of the flower.
- Showy, insect-pollinated flowers.
- Some members produce toxic alkaloids; others produce bitter or acrid
  compounds with medicinal properties.

Rosaceae: Rose Family

- Flowers with 5 petals, 5 sepals, and numerous stamens.
- Flowers typically insect-pollinated.
- Leaves are alternate, can be simple or compound.
- Leaves and/or leaflets are often oval with serrated margins and leaf-like
  stipules at their base.

Solanaceae: Nightshade Family

- Star-shaped, bisexual flowers with 5 typically fused sepals, 5 petals fused for
  less than half their length, and 5 stamens attached to the base of the corolla.
- Flowers insect-pollinated; fruit is a berry with numerous small seeds.
- Alternated leaves, compound or simple, often with an acrid smell.
- Many members of this family contain toxic or narcotic alkaloids (tobacco);
  others produce important edible fruits (tomato) or tubers (potato).

Vitaceae: Grape Family

- Climbing vines (lianas) with tendrils that coil or adhesive disks
  at their tips.
- Leaves are alternate, deeply lobed or compound.
- Tendrils and flower clusters are produced opposite the leaves.
- Flowers typically insect-pollinated.
- Clusters of fleshy berries contain relatively few seeds.

GLOSSARY*

Achene A single-seeded, dry fruit that remains closed at maturity (e.g., the "seeds" of buttercups, cinquefoils, and clematis).

Actinobacteria Filamentous, gram-positive bacteria; members of the genus *Frankia* form nitrogen-fixing root nodules in symbiosis with a variety of woody dicotyledons (e.g., alder, autumn olive). *Contrast* Rhizobium

Adventitious bud or shoot A bud or shoot that arises from any region of the plant other than the leaf axil, as from a root, a stem, or a leaf.

Adventitious root A root that originates from stem or leaf tissue rather than from another root.

Adventive Refers to an introduced or nonnative species with only a limited or temporary distribution in a given area; not widespread. *Contrast* Escape; Invasive; Naturalized; Spontaneous

Allelopathy Release of a chemical compound by one plant that inhibits the growth of another plant.

Alternate Refers to leaves or other organs arranged singly at a node. *Contrast* Opposite; Whorled

Angiosperm A plant that produces flowers with ovules enclosed in an ovary. *Contrast* Gymnosperm

Annual A plant that completes its life cycle in 1 year, including germination, flowering, seed set, and death. *See* Summer annual; Winter annual; *Contrast* Biennial

Anther Enlarged, terminal portion of the stamen that produces pollen.

---

*Definitions are based on a number of sources. The most significant are the following:

Gleason, H. A., and A. Cronquist. 1991. *Manual of Vascular Plants of Northeastern United States and Adjacent Canada*, 2nd ed. Bronx: New York Botanical Garden.
Uva, R. H., J. C. Neal, and J. M. DiTomaso. 1997. *Weeds of the Northeast.* Ithaca: Cornell University Press.
Hickey, M., and C. King. 2000. *The Cambridge Illustrated Glossary of Botanical Terms,* Cambridge: Cambridge University Press.
DiTomaso, J. M., and E. A. Healy. 2007. *Weeds of California and Other Western States.* Berkeley: University of California Press.

Apical Located at the tip of an organ, such as a leaf, root, or shoot. *Contrast* Basal; Lateral

Apomixis Production of seeds without fertilization; a form of asexual reproduction.

Archaeophyte A plant introduced into European agriculture prior to AD 1500. *Contrast* Neophyte

Aril A fleshy, often brightly colored covering on some seeds (e.g., those of the yew tree). *See* Elaiosome

Asexual reproduction *See* Vegetative reproduction

Auricle In grasses, a small, projecting lobe or ear-like appendage located where the blade meets the sheath; in broadleaf plants, an ear-like lobe that protrudes from the base of the leaf blade.

Awn Slender bristle on a grass floret.

Axil Angle formed by the upper side of a leaf and the stem to which it is attached. The position on a stem above the point of attachment of a leaf where a bud is located is the "leaf axil." *See* Subtend

Basal Located at the base of an organ, such as a stem, leaf, or flower. *Contrast* Apical; Lateral

Berry A fleshy, indehiscent fruit with 1 or more seeds embedded in pulp.

Biennial A plant that requires 2 years to complete its life cycle. During the first season the seed germinates and produces a rosette of leaves; the following year it flowers, sets seed, and dies. *Contrast* Annual

Bipinnate Refers to leaves that are branched twice, with leaflets on the second-order branches. *See* Compound leaf; *Contrast* Pinnate

Bisexual A flower having both male and female parts. *See* Perfect flower; *Contrast* Unisexual

Blade Expanded, flattened portion of a leaf; located above the petiole in broadleaf plants and above the sheath in grasses.

Bloom A waxy powder that covers the surface of a leaf or fruit, making it appear bluish or whitish. *See* Glaucous

Bolt To produce erect, flowering stems from a basal rosette of leaves; typically associated with annual or biennial species.

Bract A reduced, leaf-like structure located below a flower or inflorescence.

Bud A young shoot protected by scale leaves from which flowers or leaves develop; typically located in the axil of a leaf or bract.

Bulb A short underground shoot with leaves modified to act as food storage organs.

Bulbil A small, bulb-like structure, usually found in the angle between a leaf and stem or in place of flowers with the capacity to form a new plant.

Bur A fruiting structure covered with spines or prickles, typically dispersed as a unit.

C_3, C_4, CAM photosynthesis *See* Photosynthesis

Calyx Collectively, the sepals of a flower. *Contrast* Corolla

Capitulum A head of sessile flowers surrounded by a ring of bracts; the inflorescence characteristic of the family Asteraceae.

Capsule A dry fruit that splits open at maturity to release its seeds.

Carbohydrate An organic compound composed of carbon, hydrogen, and oxygen, such as a sugar or starch; one of the products of photosynthesis.

Carpel Female organ at the center of a flower, consisting of an ovary, a style, and a stigma. *See* Pistil

Catkin A pendulous, spike-like inflorescence of unisexual flowers that lack petals; found in the family Betulaceae.

Circumboreal Occurring all the way around the North Pole, encompassing Europe, Asia, and North America.

Clasping Refers to the base of a leaf blade that surrounds the stem to which it is attached.

Cleistogamous flower A self-pollinating flower that produces seeds without opening, as in violets and jewelweed.

Clone In plants, a group of genetically identical individuals produced by means of vegetative reproduction from a single parent.

Collar Outer side of a grass leaf located at the junction of the blade and sheath; also the base of a tree or shrub where new shoots originate. *See* Crown

Colonizer *See* Pioneer

Composite flower head Dense inflorescence of the family Asteraceae composed of individual florets carried on a receptacle subtended by bracts.

Compound leaf A leaf composed of 2 or more leaflets. Once-compound leaves have leaflets arranged along an unbranched petiole or rachis; twice-compound leaves have leaflets arranged along a branched petiole or rachis. *See* Bipinnate; Palmate; Pinnate

Cone Organ of sexual reproduction in gymnosperms; woody female cones produce seeds and transitory male cones produce pollen. *Contrast* Flower

Conical Cone-shaped.

Conifer A seed-bearing plant that produces cones; part of the gymnosperm group.

Cool-season grass Any species of grass that grows best in cool spring or fall weather and utilizes C_3 photosynthesis. *Contrast* Warm-season grass

Cordate Heart-shaped.

Corm An underground storage organ consisting of a short, vertical stem surrounded by dry, papery leaf bases.

Corolla Collective term for the petals of a flower, either separate or fused. *Contrast* Calyx

Corymb An inflorescence with branches arising at different points on a stem but reaching more or less the same height at maturity, producing a flat-topped appearance. *Contrast* Cyme; Umbel

Cotyledon Embryonic leaf of a seed-bearing plant, one or more of which are the first leaves produced by a germinating seed. Monocotyledon have one, dicotyledons have two, and gymnosperms have two or more.

Crown Base of a perennial plant, located at or just below ground level, where new shoots originate. *See* Collar; Stump sprout. Secondary use: the canopy of a tree.

Culm A jointed stem, especially the flowering stem of grasses.

Cultivar An individual perennial plant or strain of annual plant that has been selected and propagated for its unique characteristics. The word is a contraction of the phrase "cultivated variety."

Cyme A flat-topped inflorescence in which the central flowers on the main axis and the lateral branches open before the flowers below it. *Contrast* Corymb; Umbel

Deciduous Dropped at senescence, as leaves in autumn or petals after flowering.

Decussate Arranged on a stem in opposite pairs, each perpendicular to the pair above and below, forming an X shape.

Dehiscent Splitting open at maturity, as a pod releasing its seeds.

Deltoid Shaped more or less like an equilateral triangle.

Dicot Shortened term for dicotyledon.

Dicotyledon A broadleaf flowering plant (i.e., an angiosperm) that is characterized by seedlings with 2 cotyledons (along with a number of other features). *Contrast* Monocotyledon

Dimorphic Producing two different forms of the same organ, as the leaves of perennial vines.

Dioecious Producing male and female flowers on different plants of the same species. *Contrast* Monoecious

Diploid Having two sets of chromosomes in each cell nucleus. *Contrast* Haploid; Polyploid

Disk floret Central, tubular flower of some members of the family Asteraceae with both male and female organs. *Contrast* Ray floret

Dissected Divided into many slender, irregular segments but not compound; typically used to describe leaves or petals.

Doctrine of Signatures A theory dating from the time of Dioscorides stating that herbs resembling various parts of the body can be used to treat ailments of those body parts.

Dormant Resting or inactive; used to describe seeds or buds that will sprout only after experiencing a period of chilling.

Double flower A mutation that results in flowers having more petals than normal.

Drupe   A fleshy fruit containing a single seed enclosed in a hardened ovary wall (e.g., a cherry or plum).

Elaiosome   An outgrowth on the seeds of some plants containing large, oil-filled cells; typically associated with dispersal by ants (e.g., greater celandine).

Elliptical   Football-shaped; used to describe a leaf that is widest in the middle and narrowing equally toward the ends.

Endemic   A plant with a natural distribution restricted to a particular country or region.

Endophyte   A bacterium or fungus that lives within a plant (typically the leaves) for at least part of its life cycle without causing a disease.

Entire   Unbroken edge of a leaf or petal with a continuous, untoothed margin.

Epigenetics   Study of changes in organisms caused by modifications in the expression of genes rather than by alterations of the genetic code itself. Epigenesis is often associated with changes in *phenotype*.

Escape   Any cultivated plant that, under its own power, has spread outside the garden or field in which it was originally planted. *Contrast* Adventive; Invasive; Naturalized; Spontaneous; Volunteer

Exotic   Non-native; originating in a foreign country or region. *Contrast* Native

Fastigiate   Refers to the growth habit of a tree in which the branches are all more or less erect or ascending, as a Lombardy poplar. *Contrast* Weeping

Fertilization   In plants, the fusion of an ovule with a pollen cell that leads to the production of a viable seed.

Filament   Stalk of a stamen.

Floret   A little flower; an individual flower in a flower cluster, as in a grass spikelet or the composite flower head in the family Asteraceae.

Flower   Organ of sexual reproduction in angiosperms; typically composed of sepals, petals, stamens, carpels. *Contrast* Cone

Flower head   *See* Composite flower head

Frond   Leaf of a fern.

Fruit   A mature ovary of a plant including the enclosed seeds.

Gametophyte   Sexual stage in the life cycle of a plant in which the cells have half the usual number of chromosomes.

Gemma   An adventitious bud arising on a moss, liverwort, or fern leaf that can develop into an independent plant.

Genotype   Genetic makeup of an individual organism. *Contrast* Phenotype

Germination   Sprouting of a seed to produce a seedling; the production of a pollen tube by a pollen grain.

Glabrous   Smooth, without hairs. *Contrast* Pubescent

Glaucous   Covered with a waxy, bluish-green coating; used to describe leaves or fruits.

Gymnosperm A group of seed plants whose ovules are "naked," that is, not enclosed in an ovary. *Contrast* Angiosperm

Habitat Environmental conditions in which a plant grows.

Haploid Having a single set of chromosomes in each nucleus; the typical condition of egg and pollen cells. *Contrast* Diploid; Polyploid

Herb A plant with stems that die to the ground at the end of the growing season.

Herbaceous Refers to an annual plant in which both the stems and the roots are composed of non-woody tissue or a perennial plant where only aboveground stems are non-woody.

Hybrid A plant produced by the sexual crossing of parents belonging to two genetically distinct taxonomic groups.

Hypocotyl On a seedling, the region of the stem between the cotyledons and the primary root.

Indehiscent Refers to a fruit that remains closed at maturity.

Indigenous *See* Native

Inflorescence Grouping or arrangement of flowers on a stem; a cluster of flowers.

Internode Section of a stem between two adjacent nodes.

Invasive Refers to a non-native species with the capacity to proliferate and spread aggressively into natural habitats or minimally managed landscapes, often displacing native species and reducing biodiversity. *Contrast* Adventive; Escape; Naturalized; Spontaneous; Volunteer

Lanceolate More or less lance-shaped; that is, much longer than wide with a rounded base and a pointed tip.

Lateral Located on the side of an organ; typically used to describe the position of branches, roots, or flowers. *Contrast* Apical; Basal

Leaf A lateral outgrowth from the stem, typically consisting of a petiole and a blade.

Leaflet Leaf-like subunit of a compound leaf lacking an associated bud.

Legume A simple, dry fruit that opens lengthwise along two seams and is characteristic of the family Fabaceae, for example, a string bean or a pea pod.

Lenticle A small pore or opening in the bark that allows gases to pass in and out.

Liana A woody vine with flexible stems that climbs—and often smothers—trees in order to achieve greater access to sunlight.

Ligulate floret Refers to a flower type in the family Asteraceae with a long, petal-like corolla. *Contrast* Ray floret

Ligule Thin, dry membrane that projects from the top of the leaf sheath in grasses and sedges.

Lobe A rounded segment of a leaf or flower part that is larger than a tooth, typically with the adjoining sinuses extending less than halfway to the midrib; leaf or petal margins are often described as "lobed."

Margin Outer edge of a leaf or petal. *See* Entire; Lobe; Serrate

Membranous Thin, flexible, and transparent.

Monocot Shortened term for monocotyledon.

Monocotyledon A grass-like flowering plant (i.e., angiosperm) that is characterized by seedlings with 1 cotyledon (along with a number of other features). *Contrast* Dicotyledon

Monoecious Producing separate male and female flowers on the same plant. *Contrast* Dioecious

Mycorrhizae A mutually beneficial, symbiotic association between various fungi and the roots of plants that involves the exchange of minerals and carbohydrates.

Native Occurring naturally in a given region; not introduced into an area as a result of human activity; indigenous. *Contrast* Exotic

Naturalized Refers to an introduced or non-native species that reproduces on its own and is well established in a region. *Contrast* Adventive; Escape; Invasive; Spontaneous

Nectary An organ, usually located on a flower but sometimes found on leaves, that produces nectar to attract animals.

Neophyte A plant introduced into European agriculture—mainly from Asia and the Americas—after AD 1500. *Contrast* Archaeophyte

Nitrogen fixation A biochemical process carried out by bacteria in which nitrogen gas in the air is converted into ammonium that is then used to produce amino acids and proteins. *See* Root nodule; Symbiosis

Node The place on a stem where a leaf and its associated bud are attached. *See* Internode

Novel ecosystem A habitat that has been irreversibly altered by human activities and is inhabited by a cosmopolitan assemblage of organisms not found in native ecosystems.

Oblong Much longer than wide, with parallel sides and a more or less rectangular shape.

Ocrea A papery sheath that encloses the stem at the nodes formed by the fusion of 2 stipules; found in members of the family Polygonaceae.

Opposite Refers to leaves or other organs arranged directly across from each other at the same node. *Contrast* Alternate; Whorled

Ovary Part of the female flower in angiosperms that contains ovules and develops into a fruit.

Ovate Shaped like a chicken egg, with the larger end closer to the base than the tip.

Ovule A structure within the ovary that, after fertilization, develops into a seed.

Palmate Divided to the base into separate leaflets, all of which arise from the same point at the end of the petiole.

Panicle An inflorescence with a main axis and secondary branches; usually broadest at the base and tapering upward.

Pappus In the family Asteraceae, the tuft of hairs or dry scales on a seed that facilitate wind or animal dispersal.

Pedicel Stalk of a single flower or fruit within an inflorescence.

Peduncle Stalk of an inflorescence or a single flower or fruit.

Perennial Refers to a plant that lives for more than 2 years. *Contrast* Annual

Perfect flower A flower with both male and female organs. *See* Bisexual

Petal An individual part of the corolla, usually colored.

Petiole Stalk of a leaf.

pH Scale used to measure the acidity (pH < 7) or alkalinity (pH > 7) of a solution.

Phenotype An observable characteristic or trait of an organism; it can be the product of either genetic, epigenetic, or environmental factors or their interaction. *Contrast* Genotype

Photosynthesis Process whereby a green plant converts carbon dioxide and water into sugars and oxygen in the presence of sunlight. Plants have evolved three different photosynthetic pathways: C_3 **photosynthesis** is ancestral and turns CO_2 into 3-carbon compounds during the day—about 90% of plants use this pathway; C_4 **photosynthesis** involves temporarily storing CO_2 in 4-carbon compounds for later conversion to sugar—many grasses and other plants adapted to warm growing conditions use this pathway; and **Crassulacean acid metabolism** (**CAM**) takes up CO_2 at night when the stomata are open and metabolizes it during the day when the stomata are closed—utilized by cacti and other drought-adapted species.

Phytoremediation Use of plants to clean up land contaminated by a variety of human-generated waste products, such as heavy metals, petroleum products, acid mine drainage, and excessive amounts of dissolved nitrogen and phosphorus.

Pinnate Refers to a compound leaf with leaflets arranged along both sides of a petiole; they may be odd-pinnate with a single terminal leaflet or even-pinnate with a terminal pair of leaflets. *Contrast* Bipinnate

Pioneer A plant that colonizes bare ground; an early successional species. *Contrast* Invasive; Weed

Pistil Female organ of a flower, consisting of an ovary, style, and stigma. *See* Carpel

Pistillate Refers to flowers having only female organs.

Pith Central, spongy tissue in a stem, twig, or root that is surrounded by a cylinder of woody tissue.

Pollen Powdery mass shed from anthers of a flower that contains the male reproductive cells.

Pollination Transfer of pollen from an anther to a stigma, which, if successful, results in fertilization and the production of a seed.

Polyploid Having more than one set of chromosomes in each cell nucleus. *Contrast* Diploid; Haploid

Preadaptation An anatomical or physiological trait that evolved under one set of ecological conditions and, by chance, proves advantageous under a completely different set of circumstances.

Prostrate Lying flat on the ground; used to describe the growth habit of low-growing plants.

Pubescent Covered with short hairs; typically used to describe leaves and stems. *Contrast* Glabrous

Raceme An elongated inflorescence with stalked flowers on an unbranched central axis; flowers develop from the bottom up. *Contrast* Spike

Rachis Central axis of a pinnately compound leaf.

Ray floret One of the outer, irregular flowers in the flower heads of some plants in the family Asteraceae that produce a single, strap-like "petal." *Contrast* Disk floret

Receptacle Basal part of a flower to which the other flower parts are attached; it is sometimes enlarged and fleshy, as in a strawberry or blackberry.

Recurved Bent or curved downward or backward.

Revolute Rolled backward at the margin such that the upper surface of a leaf is exposed and the lower surface is hidden.

Rhizobium Rod-shaped bacteria that are capable of nitrogen fixation in symbiosis with plants in the family Fabaceae. *Contrast* Actinobacteria

Rhizome A creeping, underground stem that produces new shoots and adventitious roots. *Contrast* Stolon

Root Lower portion of the plant's axis that anchors it in the soil and absorbs nutrients and water. The primary root develops from the embryo, and secondary roots are branches off the primary root.

Root nodule An outgrowth on the roots of certain plants that contains *Rhizobium* or *Frankia* bacteria and is the site of symbiotic nitrogen fixation.

Root sucker A shoot that arises from an adventitious bud on a root.

Rosette In herbaceous plants, a circular cluster of leaves located at ground level. The stem is compressed and the leaves are separated by very short internodes.

Ruderal Refers to a plant that grows in waste places or requires soil disturbance to become established. From Latin *rudera*, "ruins" or "rubbish," plural of *rudus*, "broken stone." In botanical parlance, a disturbance-adapted species.

Rugose Having a wrinkled surface; in leaves, depressed along the veins and elevated between them.

Runner *See* Stolon

Samara A dry, indehiscent winged fruit, such as maple or ash seeds.

Scale A reduced or rudimentary leaf; usually surrounding a dormant bud.

Scandent Weakly climbing, as in shrub roses.

Scion In grafting, a young shoot with special characteristics that is spliced or grafted onto a rooted understock.

Seed A ripened ovule consisting of a protective coat enclosing an embryo and food reserves; the product of sexual reproduction.

Seed bank The totality of viable seeds that lie buried in the soil at any point in time; they typically germinate following some disturbance that brings them to the soil surface.

Seed head An inflorescence bearing mature fruit; especially in the family Asteraceae.

Seed leaf *See* Cotyledon

Sepal In a flower, the outermost whorl of leaf-like, usually green appendages below the petals. *See* Calyx

Serrate Refers to the margin of a leaf with sharp, forward-pointing teeth. *Contrast* Entire; Lobe

Sessile Attached directly by the base, as a leaf without a petiole or a flower without a pedicel.

Sexual reproduction Production of a plant from seed. *Contrast* Vegetative reproduction

Sheath In grasses, the portion of a leaf below the blade that encloses the stem and the emerging new leaves.

Shoot A young green stem with leaves, flowers, or both.

Shrub A woody plant that, when undisturbed, branches spontaneously at or below ground level to produce multiple trunks. *Contrast* Tree

Simple leaf A leaf blade that consists of a single piece but may be deeply lobed or divided. *Contrast* Compound leaf

Sinuses Indentations between the teeth or lobes on the margin of a leaf.

Species A population of individuals sharing common morphological features and evolutionary history; the basic unit of classification in the nomenclatural hierarchy.

Spike An elongated inflorescence with sessile flowers on an unbranched central axis. *Contrast* Raceme

Spikelet In grasses and sedges, an inflorescence consisting of 1 to many flowers subtended by minute bracts.

Spontaneous Refers to a plant that grows in an area without being cultivated by humans; it may be either a native or an introduced species. *Contrast* Adventive; Escape; Invasive; Naturalized

Spore Tiny, dust-like reproductive unit of mosses, ferns, and horsetails.

Sporophyte Mature stage in the life cycle of mosses, ferns, and horsetails that produces spores.

Stamen Male organ of a flower, consisting of a stalk (filament) and a pollen-producing organ (anther).

Staminate Flowers that have only male organs.

Stigma Apex of the style, usually enlarged and sticky, on which the pollen grains land and germinate. *See* Carpel; Pistil

Stipule A small, leafy outgrowth at the base of a petiole; typically found in pairs.

Stolon A stem that grows horizontally along the surface of the ground, producing roots at its nodes and new plants from its buds. *Contrast* Rhizome

Stomata Small pores typically located on the undersides of leaves which, when they are open, allow gasses to pass in and out of the plant.

Strobilus Cone-like reproductive structures of horsetails and conifers.

Stump sprout A shoot that emerges from the base of a tree or shrub, usually after some form of traumatic injury, such as logging; a basal sprout. *See* Crown

Style Elongated portion of a carpel, above the ovary, with the stigma at its tip.

Subtend To be located beneath a given structure, as a leaf at the base of a bud. *See* Axil

Succession Changes in the composition of a plant association over time, typically initiated by some form of disturbance. **Early succession** refers to the beginning stages of the process, immediately following a disturbance event; **late succession** refers to the latter stages of the process when the composition of the vegetation has stabilized.

Summer annual A plant that germinates in late spring or early summer, grows best in warm weather, and completes its life cycle before winter. *Contrast* Winter annual

Symbiosis A long-lasting association between two species of organisms that may or may not be mutually beneficial; literally the word means "living together."

Taproot A thick, downward-growing root with few side branches; adapted to penetrating heavy soils and storing carbohydrates.

Tendril A twining, thread-like structure produced by the stem or leaf of a climbing plant that enables it to cling to an object.

Terminal Located at the tip or apex (e.g., a terminal inflorescence). *Contrast* Lateral

Tiller A lateral shoot that emerges from the base of a grass or other monocot.

Tree A plant that, when undisturbed, develops a single woody trunk. *Contrast* Shrub

Trifoliate A compound leaf consisting of 3 leaflets, as in clover or poison ivy.

Truncate Ending abruptly as if cut off across the base or tip; flat.

Tuber Swollen tip of an underground stem or rhizome; an underground food storage organ that sprouts when growing conditions are favorable (e.g., a potato).

Tubular floret *See* Disk floret

Umbel A cluster of flowers—typically flat-topped—with numerous pedicels arising from a common point. *Contrast* Corymb; Cyme

Understock In grafting, the rooted plant (usually a seedling) to which the scion is spliced or grafted.

Undulate Having a wavy, up-and-down margin; used to describe leaves or petals.

Unisexual A flower having organs of only 1 sex. *Contrast* Bisexual

Variegated Having two or more colors.

Vegetative reproduction Asexual reproduction by any means other than from seed, including bulbs, rhizomes, stump sprouts, root suckers, and leaves. *See* Clone; *Contrast* Sexual reproduction

Vein An externally visible strand of vascular tissue in a leaf or other flat organ.

Volunteer *See* Escape

Warm-season grass Any species of grass that grows best in warm summer weather and utilizes C_4 photosynthesis. *Contrast* Cool-season grass

Weed A plant that grows rapidly and abundantly in a place where it is not wanted; a plant that aggressively colonizes disturbed habitats. *Contrast* Invasive

Weeping With branches bending over or hanging down, as a weeping willow. *Contrast* Fastigiate

Whorled Having a ring of 3 or more leaves or other organs radiating from a node. *Contrast* Alternate; Opposite

Wing A thin, flat extension or projection from the side of a stem or leaf.

Winter annual A plant that germinates in fall or winter and completes its life cycle the following spring. Winter annuals are tolerant of cold weather, overwinter as rosettes, and grow best in cool, moist conditions. *Contrast* Summer annual

REFERENCES

Alberti, M., C. C. Correa, J. M. Marzluff, A. P. Hendry, E. P. Palkovacs, K. M. Gotanda, V. M. Hunta, T. M. Apgar, and Y. Zhou. 2017. Global urban signatures of phenotypic change in animal and plant populations. *Proceedings of the National Academy of Sciences*. doi:10.101/201606034.

Anderson, E. 1952. *Plants, Man and Life*. Boston: Little, Brown.

Anderson, E., and R. E. Woodson. 1935. The species of *Tradescantia* indigenous to the United States. *Contributions from the Arnold Arboretum of Harvard University*, no. IX.

Arnold, C. L., and C. J. Gibbons. 1996. Impervious surface coverage: the emergence of a key environmental indicator. *Journal of the American Planning Association* 62(2):243–258.

Aronson, M. F. J., S. N. Handel, I. P. LaPuma, and S. E. Clemants. 2015. Urbanization promotes non-native woody species and diverse plant assemblages in the New York metropolitan region. *Urban Ecosystems* 18:31–45.

Baker, H. G. 1974. The evolution of weeds. *Annual Review of Ecology and Systematics* 5:1–24.

Barnes, B. V., and W. H. Wagner Jr. 2004. *Michigan Trees*. Ann Arbor: University of Michigan Press.

Barney, J. N. 2006. North American history of two invasive plant species: phytogeographic distribution, dispersal vectors and multiple introductions. *Biological Invasions* 8:703–717.

Barton, J., and J. N. Pretty. 2010. Urban ecology and human health and wellness. In *Urban Ecology*, K. J. Gaston (ed.), pp. 202–229. Cambridge: Cambridge University Press.

Bartram, J. 1992. *The Correspondence of John Bartram, 1734–1777*, edited by E. Berkeley and D. Smith Berkeley. Gainesville: University Press of Florida.

Bormann, F. H., D. Balmori, and G. T. Geballe. 1993. *Redesigning the American Lawn: A Search for Environmental Harmony*. New Haven: Yale University Press.

Brauer, J., and M. A. Gerber. 2002. Population differentiation in the range expansion of a native maritime plant, *Solidago sempervirens* L. *International Journal of Plant Sciences* 163(1):141–150.

Burkholder, S. 2012. The new ecology of vacancy: rethinking land use in shrinking cities. *Sustainability* 4:1154–1172.

Byrne, L. B. 2007. Habitat structure: a fundamental concept and framework for urban soil ecology. *Urban Ecosystems* 10:255–274.

Carroll, S. P. 2011. Conciliation biology: the eco-evolutionary management of permanently invaded biotic systems. *Evolutionary Applications* 4:184–199.

Cheptou, P.-O., O. Carrue, S. Rouifed, and A. Cantarel. 2008. Rapid evolution of seed dispersal in an urban environment in the weed *Crepis sancta*. *PNAS* 105(10):3796–3799.

Clemants, S. E., and G. Moore. 2005. The changing flora of the New York metropolitan region. *Urban Habitats* 3:192–210.

Coates, P. 2006. *Strangers on the Land: American Perceptions of Immigrant and Invasive Species*. Berkeley: University of California Press.

Crèvecœur, J. Hector St John de. 1783 [reprint 1997]. *Letters from an American Farmer*. Oxford: Oxford University Press

Culley, T. M., and N. A. Hardiman. 2007. The beginning of a new invasive plant: a history of the ornamental Callery pear in the United States. *BioScience* 57(11):956–964.

Darlington, W., and G. Thurber. 1859. *American Weeds and Useful Plants*. New York: A. O. Moore.

Davis, M., M. K. Chew, R. J. Hobbs, A. E. Lugo, J. J. Ewel, G. J. Vermeij, J. H. Brown, M. L. Rosenzweig, M. R. Gardener, S. P. Carroll, K. Thompson, S. T. A. Pickett, J. C. Stromberg, P. Del Tredici, K. N. Suding, J. G. Ehrenfeld, J. P. Grime, J. Mascoaro, and J. C. Briggs. 2011. Don't judge species on their origins. *Nature* 474:153–154.

Decina, S. M., P. H. Templer, L. R. Hutyra, C. K. Gately, and P. Rao. 2017. Variability, drivers, and effects of atmospheric nitrogen inputs across an urban area: emerging patterns among human activities, the atmosphere, and soils. *Science of the Total Environment* 609:1524–1534.

DeGasperis, B. G., and G. Motzkin. 2007. Windows of opportunity: historical and ecological controls on *Berberis thunbergii* invasions. *Ecology* 88(12):3115–3125.

Del Tredici, P. 1977. The buried seeds of *Comptonia peregrina*, the sweet fern. *Bulletin of the Torrey Botanical Club* 104(3):270–275.

Del Tredici, P. 1986. The great catalpa craze. *Arnoldia* 46(2):3–10.

Del Tredici, P. 2001. Sprouting in temperate trees: a morphological and ecological review. *Botanical Review* 67(2):121–140.

Del Tredici, P. 2007. The role of horticulture in a changing world. In *Botanical Progress, Horticultural Innovation, and Cultural Changes*, M. Conan and W. J. Kress (eds.), pp. 259–264. Washington, D.C.: Dumbarton Oaks.

Del Tredici, P. 2010. Spontaneous urban vegetation: reflections of change in a globalized world. *Nature and Culture* 5(3):299–315.

Del Tredici, P. 2014. Untangling the twisted tale of oriental bittersweet. *Arnoldia* 71(3):2–18.

Del Tredici, P. 2017a. The introduction of Japanese plants into North America. *Botanical Review* 83(3):215–252.

Del Tredici, P. 2017b. The introduction of Japanese knotweed, *Reynoutria japonica*, into North America. *Journal of the Torrey Botanical Society* 144(3):406–416.

Del Tredici, P., and M. Luegering. 2014. A cosmopolitan urban meadow for the northeast. *Harvard Design Magazine* 37 (http://www.harvarddesignmagazine.org/issues/37/a-cosmopolitan-urban-meadow-for-the-northeast).

Del Tredici, P., and T. Rueb. 2017. Other order: sound walk for an urban wild. *Arnoldia* 75(1):14–25.

Donihue, C. M., and M. R. Lambert. 2014. Adaptive evolution in urban ecosystems. *Ambio.* doi:10.1007/s13280-014-0547-2.

Duffy, G. A., and S. L. Chown. 2016. Urban warming favors C_4 plants in temperate European cities. *Journal of Ecology* 104:1618–1626.

Dunnett, N., and J. Hitchmough, eds. 2004. *The Dynamic Landscape: Design, Ecology and Management of Naturalistic Urban Planting.* London: Spon Press.

Ellis, E. C., J. O. Ellisa, J. O. Kapla, D. Q. Fuller, S. Vavrus, K. K. Goldewijk, and P. H. Verburg. 2013. Used planet: a global history. *PNAS* 110(20):7978–7985.

Elton, C. S. 1958. *The Ecology of Invasions by Animals and Plants.* London: Methuen.

Emerson, R. W. 1879. *Fortune of the Republic.* Cambridge: Riverside Press.

Fogg, J. M. Jr. 1945. *Weeds of Lawn and Garden.* Philadelphia: University of Pennsylvania Press.

Fridley, J. D. 2013. Plant invasions across the northern hemisphere: a deep-time perspective. *Annals of the New York Academy of Sciences* 1293:8–17.

Fridley, J. D., and D. F. Sax. 2014. The imbalance of nature: revisiting a Darwinian framework for invasion biology. *Global Ecology and Biogeography* 23:1157–1166.

Garvin, E., C. Branas, S. Keddem, and J. Sellman. 2012. More than just an eyesore: local insights and solutions on vacant land and urban health. *Journal of Urban Health* 90:412–426.

George, K., L. H. Ziska, J. A. Bunce, and B. Quebedeaux. 2007. Elevated atmospheric CO_2 concentration and temperature across an urban-rural transect. *Atmospheric Environment* 41:7654–7665.

Gilbert, O. L. 1989. *The Ecology of Urban Habitats.* London: Chapman and Hall.

Gleason, H. A., and A. Cronquist. 1991. *Manual of Vascular Plants of Northeastern United States and Adjacent Canada*, 2nd ed. New York: New York Botanical Garden.

Godefroid, S., D. Monbaliu, and N. Koedam. 2007. The role of soil and micro-climatic variables in the distribution patterns of urban wasteland flora in Brussels, Belgium. *Landscape and Urban Planning* 80:45–55.

Gould, S. J. 1998. An evolutionary perspective on strengths, fallacies, and confusions in the concept of native plants. *Arnoldia* 58(1):2–10.

Gregg, J. W., C. G. Jones, and T. E. Dawson. 2003. Urbanization effects on tree growth in the vicinity of New York City. *Nature* 424:183–187.

Grime, J. P. 2001. *Plant Strategies, Vegetation Processes, and Ecosystem Properties*, 2nd ed. New York: John Wiley and Sons.

Gulachenski, A., B. M. Ghersi, A. E. Lesen, and M. J. Blum. 2016. Abandonment, ecological assembly and public health risks in counter-urbanizing cities. *Sustainability* 8:491. doi:10.3390/su8050491.

Gunther, R. T. (ed.). 1959. *The Greek Herbal of Dioscorides: Illustrated by a Byzantine, A.D. 512. Englished by John Goodyer, A.D. 1655.* New York: Hafner.

Hauser, S. C. 1996. *Nature's Revenge.* New York: Lyons and Burford.

Heiser, C. B. 2003. *Weeds in My Garden.* Portland: Timber Press.

Hickey, M., and C. King. 2000. *The Cambridge Illustrated Glossary of Botanical Terms.* Cambridge: Cambridge University Press.

Hobbs, R. J., E. S. Higgs, and C. M. Hall (eds.). 2013. *Novel Ecosystems: Intervening in the New Ecological Order.* West Sussex: Wiley-Blackwell.

Hofmeister, S. 2009. Nature running wild: a social-ecological perspective on wilderness. *Nature and Culture* 4(3):293–315.

Hu, S.-Y. 1979. *Ailanthus. Arnoldia* 39(2):29–50.

Johnson, M. T. J., and J. Munshi-South. 2017. Evolution of life in urban environments. *Science* 358, eaam8327. doi:10.1126/science.aam8327.

Johnson, M. T. J., C. M. Prashad, M. Lavoignat, and H. S. Saini. 2018. Contrasting the effects of natural selection, genetic drift and gene flow on urban evolution in white clover (*Trifolium repens*). *Proceedings of the Royal Society Part B* 285(1883). doi:10.1098/rspb.2018.1019.

Jorgensen, A., and R. Keenan (eds.). 2012. *Urban Wildscapes.* London: Routledge.

Josselyn, J. 1672 [reprint 1865]. *New-England's Rarities Discovered*, E. Tuckerman (ed.). Boston: William Veazie.

Kareiva, P., S. Watts, R. McDonald, and T. Boucher. 2007. Domesticated nature: shaping landscapes and ecosystems for human welfare. *Science* 316:1866–1869.

Keil, A. 2005. Use and perception of post-industrial urban landscapes in the Ruhr. In *Wild Urban Woodlands*, I. Kowarik and S. Körner (eds.), pp. 117–130. Berlin: Springer.

Kelcey, J. G., and N. Müller (eds.). 2011. *Plants and Habitats of European Cities.* Berlin: Springer.

Kenfield, W. G. (alias F. E. Egler). 1966. *The Wild Gardener in the Wild Landscape.* New York: Hafner.

Knapp, S., S Kühn, J. Stolle, and S. Klotz. 2010. Changes in the functional composition of a central European flora over three centuries. *Perspectives in Ecology, Evolution and Systematics* 12:235–244.

Kowarik, I. 2003. Human agency in biological invasions: secondary releases foster naturalization and population expansion of alien plant species. *Biological Invasions* 5:293–312.

Kowarik, I. 2011. Novel ecosystems, biodiversity and conservation. *Environmental Pollution* 159:1974–1983.

Kowarik, I. 2018. Urban wilderness: supply, demand, and access. *Urban Forestry & Urban Greening* 28:336–347

Kowarik, I., and S. Körner (eds.). 2005. *Wild Urban Woodlands.* Berlin: Springer-Verlag.

Kowarik, I., and A. Langer. 2005. Natur-Park Südgelände: linking conservation and recreation in an abandoned railyard in Berlin. In *Wild Urban Woodlands*, I. Kowarik and S. Körner (eds.), pp. 287–299. Berlin: Springer-Verlag.

Kühn, N. 2006. Intentions for the unintentional spontaneous vegetation as the basis for innovative planting design in urban areas. *Journal of Landscape Architecture* (autumn):46–53.

Larson, D., U. Matthes, P. E. Kelly, J. Lundholm, and J. Garrath. 2004. *The Urban Cliff Revolution.* Markham: Fitzhenry and Whiteside.

Lundholm, J. T., and A. Marlin. 2006. Habitat origins and microhabitat preferences of urban plant species. *Urban Ecosystems* 9:139–159.

Lundholm, J. T., and P. J. Richardson. 2010. Habitat analogues for reconciliation ecology in urban and industrial environments. *Journal of Applied Ecology* 47:966–975.

Mack, R. N. 2000. Cultivation fosters plant naturalization by reducing environmental stochasticity. *Biological Invasions* 2:111–122.

Mack, R. N. 2003. Plant naturalizations and invasions in the eastern United States: 1634–1860. *Annals of the Missouri Botanical Garden* 90:77–90.

Mack, R. N., and M. Erneberg. 2002. The United States naturalized flora: largely the product of deliberate introductions. *Annals of the Missouri Botanical Garden* 89:176–189.

Marcotullio, P. 2011. Urban soils. In *The Routledge Handbook of Urban Ecology*, I. Douglas, D. Goode, M. C. Houck, and R. Wang (eds.), pp. 164–186. London: Routledge.

Marks, P. L. 1983. On the origin of the field plants of the northeastern United States. *American Naturalist* 122(2):210–228.

Maurer, U., T. Peschel, and S. Schmitz. 2000. The flora of selected urban land-use types in Berlin and Potsdam with regard to nature conservation in cities. *Landscape and Urban Planning* 46:209–215.

Meerts, P., T. Baya, and C. Lefèbvre. 1998. Allozyme variation in the annual weed species complex *Polygonum aviculare* (Polygonaceae) in relation to ploidy level and colonizing ability. *Plant Systematics and Evolution* 211:239–256.

Mehrhoff, L. J. 2000. Immigration and expansion of the New England flora. *Rhodora* 102:280–298.

Miller, A. B. 1976. *Shaker Herbs: A History and a Compendium*. New York: Clarkson N. Potter.

Mohan, J. E., L. H. Ziska, W. H. Schlesinger, R. B. Thomas, R. C. Sicher, K. George, and J. S. Clark. 2006. Biomass and toxicity responses of poison ivy (*Toxicodendron radicans*) to elevated atmospheric CO_2. *Proceedings of the National Academy of Sciences (USA)* 103(24):9086–9089.

Moore, G., J. Macklin, and L. DeCesare. 2010. A brief history of Asa Gray's *Manual of Botany. Harvard Papers in Botany* 15(2):277–286.

Mühlenbach, V. 1979. Contribution to the synanthropic (adventive) flora of the railroads in St. Louis, Missouri, U.S.A. *Annals of the Missouri Botanical Garden* 66(1):1–108.

Müller, N., M. Ignatieva, C. H. Nilon, P. Werner, and W. C. Zipperer. 2013. Patterns and trends in urban biodiversity and landscape design. In *Urbanization, Biodiversity and Ecosystem Services: Challenges and Opportunities: A Global Perspective*, T. Elmqvist et al. (eds.), pp. 123–174. Dordrecht: Springer. doi:10.1007/978-94-007-7088-1.

Muratet, A., N. Machon, F. Jiguet, J. Moret, and E. Porcher. 2007. The role of urban structures in the distribution of wasteland flora in the greater Paris area, France. *Ecosystems* 10(4):661–671.

Murcia C., J. Aronson, G. H. Kattan, D. Moreno-Mateos, K. Dixon, and D. Simberloff. 2014. A critique of the "novel ecosystem" concept. *Trends in Ecology and Evolution* 29(1):548–553.

Nassauer, J. I., and J. Raskin. 2014. Urban vacancy and land use legacies: a frontier for urban ecological research, design, and planning. *Landscape and Urban Planning* 125:245–253.

Nature Conservancy. 2016. *Planting Healthy Air*. Arlington: The Nature Conservancy.

Nuttall, T. 1846. *The North American Sylva*, Vol. II. Philadelphia: Townsend Ward.

Pataki, D. E., M. Carreiro, J. Cherrier, N. E. Grulke, V. Jennings, S. Pincetl, R. V. Pouyat, T. H. Whitlow, and W. C. Zipperer. 2011. Coupling biogeochemical cycles in urban environments: ecosystem services, green solutions and misconceptions. *Frontiers in Ecology and the Environment* 9:27–36.

Pauly, P. J. 2007. *Fruits and Plains: The Horticultural Transformation of America.* Cambridge: Harvard University Press.

Peterken, G. 2013. *Meadows.* Oxford: British Wildlife Publishing.

Pickett, S. T. A., M. L. Cadenasso, J. M. Grove, P. M. Groffman, L. E. Band, C. G. Boone, W. R. Burch Jr., C. S. B. Grimmond, J. Hom, J. C. Jenkins, N. L. Law, C. H. Nilon, R. V. Pouyat, K. Szlavecz, P. S. Warren, and M. A. Wilson. 2008. Beyond urban legends: an emerging framework of urban ecology, as illustrated by the Baltimore Ecosystem Study. *BioScience* 59(2):139–150.

Pickett, S. T. A., M. L. Cadenasso, J. M. Grove, C. G. Boone, P. M. Groffman, E. Irwin, S. S. Kaushal, V. Marshall, B. P. McGrath, C. H. Nilon, R.V. Pouyat, K. Szlavecz, A. Troy, and P. Warren. 2011. Urban ecological systems: scientific foundations and a decade of progress. *Journal of Environmental Management* 92:331–362.

Porębska, G., and A. Ostrowska. 1999. Heavy metal accumulation in wild plants: implications for phytoremediation. *Polish Journal of Environmental Studies* 8(6):433–442.

Pyšek, P. 1998. Alien and native species in Central European urban floras: a quantitative comparison. *Journal of Biogeography* 25:155–163.

Pyšek, P., Z. Chocholoušková, A. Pyšek, V. Jarošík, M. Chytry, and L. Tichy. 2004. Trends in species diversity and composition of urban vegetation over three decades. *Journal of Vegetation Science* 15:781–788.

Raciti, S. M., L. R. Hutyra, P. Rao, and A. C. Finzi. 2012. Inconsistent definitions of "urban" result in different conclusions about the size of urban carbon and nitrogen stocks. *Ecological Applications* 22(3):1015–1035.

Rehder, A. 1946. On the history of the introduction of woody plants into North America. *Arnoldia* 6(4–5):13–28.

Riddle, J. M. 1999. *Eve's Herbs: A History of Contraception and Abortion in the West.* Cambridge: Harvard University Press.

Riley, C. B., K. I. Perry, K. Ard, and M. M. Gardiner. 2018. Asset or liability? Ecological and sociological tradeoffs of urban spontaneous vegetation on vacant land in shrinking cities. *Sustainability* 10:2139. doi:10.3390/su10072139.

Rink, D. 2005. Surrogate nature or wilderness? Social perceptions and notions of nature in an urban context. In *Wild Urban Woodlands,* I. Kowarik and S. Körner (eds.), pp. 67–80. Berlin: Springer.

Robinson, S. L., and J. T. Lundholm. 2012. Ecosystem services provided by urban spontaneous vegetation. *Urban Ecosystems* 15:545–557.

Roy, B. A., H. M. Alexander, J. Davidson, F. Campbell, J. J. Burdon, R. Sniezko, and C. Brasier. 2014. Increasing forest loss worldwide from invasive pests require new trade regulations. *Frontiers in Ecology and the Environment* 12(8):457–465.

Sagoff, M. 2005. Do non-native species threaten the natural environment? *Journal of Argicultural and Environmental Ethics* 18:215–236.

Salisbury, E. 1961. *Weeds and Aliens.* London: Collins.

Saltonstall, K. 2002. Cryptic invasion by a non-native genotype of *Phragmites australis* into North America. *Proceedings of the National Academy of Sciences* 99(4):2445–2449.

Sargent, C. S. 1890. *Rosa multiflora. Garden and Forest* 3:404–406.

Schierenbeck, K. A, and N. C. Ellstrand. 2009. Hybridization and the evolution of invasiveness in plants and other organisms. *Biological Invasions* 11:1093–1105.

Schilthuizen, M. 2018. *Darwin Comes to Town*. New York: Picador.

Sieghardt, M., E. Mursch-Radlgruber, E. Paoletti, E. Couenberg, A. Dimitrakopou-lus, F. Rego, A. Hatzistathis, and T. B. Randrup. 2005. The abiotic urban environ-ment: impact of urban growing conditions on urban vegetation. In *Urban Forests and Trees*, C. C. Konijnendijk, K. Nilsson, T. Randrup, and J. Schipperijn (eds.), pp. 281–323. Berlin: Springer-Verlag.

Smith, B. 1943. *A Tree Grows in Brooklyn*. New York: Harper and Brothers.

Spirn, A. W. 1984. *The Granite Garden: Urban Nature and Human Design*. New York: Basic Books.

Starfinger, U., I. Kowarik, M. Rode, and H. Schepker. 2003. From desirable ornamental plant to pest to accepted addition to the flora—the perception of an alien tree species through the centuries. *Biological Invasions* 5:323–335.

Sukopp, H. 2004. Human-caused impact on preserved vegetation. *Landscape and Urban Planning* 68(4):347–355.

Tallamy, D. W. 2007. *Bringing Nature Home*. Portland: Timber Press.

Thomas, C. D. 2017. *Inheritors of the Earth*. New York: Public Affairs.

Thompson, K. 2015. *Where Do Camels Belong?* London: Profile Books.

Uva, R. H., J. C. Neal, and J. M. DiTomaso. 1997. *Weeds of the Northeast*. Ithaca: Cornell University Press.

Walther, G.-R., A. Roques, P. E. Hulme, M. T. Sykes, P. Pyšek, I. Kühn, and M. Zo-bel. 2009. Alien species in a warmer world: risks and opportunities. *Trends in Ecology and Evolution* 24(12):686–693.

Weber, F., I. Kowarik, and I. Säumel. 2014. A walk on the wild side: perceptions of roadside vegetation beyond trees. *Urban Forestry and Urban Greening* 13:205–212.

Weiss, J., W. Burghardt, P. Gausmann, R. Haag, H. Haeupler, M. Hamann, B. Leder, A. Schulte, and I. Stempelmann. 2005. Nature returns to abandoned industrial land: monitoring succession in urban-industrial woodlands in the German Ruhr. In *Wild Urban Woodlands*, I. Kowarik and S. Korner (eds.), pp. 143–162. Berlin: Springer-Verlag.

Whitney, G. G. 1985. A quantitative analysis of the flora and plant communities of a representative midwestern U.S. town. *Urban Ecology* 6:143–160.

Wittig, R. 2004. The origin and development of the urban flora of Central Europe. *Urban Ecosystems* 7:323–339.

Wittig, R., and U. Becker. 2010. The spontaneous flora around street trees in cities—A striking example for the worldwide homogenization of the flora of urban habitats. *Flora* 205:704–709.

Zerbe, S., U. Maurer, S. Schmitz, and H. Sukopp. 2003. Biodiversity in Berlin and its potential for nature conservation. *Landscape and Urban Planning* 62:139–148.

Zipperer, W. C. 2011. The process of natural succession in urban areas. In *The Routledge Handbook of Urban Ecology,* I. Douglas, D. Goode, M. C. Houck, and R. Wang (eds.), pp. 187–197. London: Routledge.

Zipperer, W. C., S. M. Siginni, and R. V. Pouyat. 1997. Urban tree cover: an ecological perspective. *Urban Ecosystems* 1:229–246.

Ziska, L. H., and J. S. Dukes. 2011. *Weed Biology and Climate Change*. New York: Wiley-Blackwell.

Ziska, L. H., D. E. Gebhard, D. A. Frenz, S. Faulkner, B. D. Singer, and J. G. Straka. 2003. Cities as harbingers of climate change: common ragweed, urbanization and public health. *Journal of Allergy and Clinical Immunology* 111(2):290–295.

Plant Identification Reference Books Consulted but Not Cited in the Text

Agricultural Research Service (USDA). 1971. *Common Weeds of the United States.* New York: Dover Publications.

Brown, L. 1977. *Weeds in Winter.* Boston: Houghton Mifflin.

Brown, L. 1979. *Grasses: An Identification Guide.* Boston: Houghton Mifflin.

Bryson, C. T., and M. S. DeFelice. 2009. *Weeds of the South.* Athens: University of Georgia Press.

Dirr, M. A. 1998. *Manual of Woody Landscape Plants*, 5th ed. Champaign: Stipes Publishing.

DiTomaso, J. M., and E. A. Healy. 2007. *Weeds of California and Other Western States.* Berkeley: University of California Press.

Eastman, J. 1992. *The Book of Forest and Thicket.* Mechanicsburg: Stackpole Books.

Eastman, J. 1995. *The Book of Swamp and Bog.* Mechanicsburg: Stackpole Books.

Eastman, J. 2003. *The Book of Field and Roadside.* Mechanicsburg: Stackpole Books.

Fernald, M. L., and A. C. Kinsey. 1943. *Edible Wild Plants of Eastern North America.* Cornwall-on-Hudson: Idlewild Press.

Fitter, R., A. Fitter, and M. Blamey. 1974. *The Wildflowers of Britain and Northern Europe.* Collins: London.

Foster, S., and J. A. Duke. 1990. *A Field Guide to Medicinal Plants of Eastern and Central North America.* Boston: Houghton Mifflin.

Hubbard, C. E. 1992. *Grasses*, 3rd ed. London: Penguin Books.

Hutchinson, J. 1946. *Common Wildflowers.* West Drayton: Penguin Books.

Hutchinson, J. 1948. *More Common Wildflowers.* West Drayton: Penguin Books.

Jones, P. 1994. *Just Weeds: History, Myths and Uses.* Shelburne: Chapters Publishing.

Kaufman, S. R., and W. Kaufman. 2007. *Invasive Plants.* Mechanicsburg: Stackpole Books.

Newcomb, L. 1977. *Newcomb's Wildflower Guide.* New York: Little, Brown.

Page, N., and R. E. Weaver. 1975. *Wild Plants in the City.* New York: Quadrangle. [Reprinted from: *Arnoldia* 34(4):137–232, 1974].

Rhoads, A. F., and T. A. Block. 2005. *Trees of Pennsylvania.* Philadelphia: University of Pennsylvania Press.

Royer, F., and R. Dickinson. 1999. *Weeds of the Northern U.S. and Canada.* Edmonton: Lone Pine Publishing and University of Alberta Press.

INDEX

Page numbers in **boldface** indicate the location of the main text description. Entries without a boldface page number are synonyms.

Acalypha
   gracilens, 230
   rhomboidea, **230**
   virginica, 230
Acer
   negundo, **112**, 367, 373
   platanoides, **114**, 374
   pseudoplatanus, **116**, 373
   rubrum, 116
   saccharinum, **118**, 367–368, 373
   saccharum, 114
Achillea millifolium, **146**, 370, 372
Ageratina altissima, **148**
Agropyron repens, 340
Ailanthus altissima, 2, **120**, 367, 368, 373
Albizia julibrissin, 368, 373
alder, black, **44**, 369, 373
alfalfa, 238, 370
allelopathy, 74, 120, 158, 198
Alliaria petiolata, 196, **198**, 368
Allium vineale, **320**
Alnus
   glutinosa, **44**, 369, 373
   incana, 44
   serrulata, 44
   viridis, 44
alyssum, hoary, **202**
Amaranthus
   blitoides, 134
   caudatus, 134
   palmeri, 134
   retroflexus, **134**
   species, 367, 368

Ambrosia artemisiifolia, 20, **150**
Amorpha fruticosa, **64**
Ampelopsis glandulosa var.
      brevipedunculata, **128**, 368
Anderson, Edgar, 1, 3
Apocynum
   androsaemifolium, 140
   cannabinum, **140**
apple
   common edible, 90, 374
   wild, **90**, 369
archaeophytes, 8–9, 19
Arctium
   lappa, 152
   minus, **152**, 370, 371
Arenaria rubra, 220
Arnold Arboretum (Boston), 42, 96,
      104
Artemesia vulgaris, **154**, 367
Asclepias syriaca, **142**, 372
ash
   white, 78, 373
   green, **78**, 368, 373
aspen, quaking, 4, **108**, 373
aster
   heart-leaved, 186
   white heath, **186**, 372
   white wood, 148, 186
Aster
   cordifolius, 186
   divaricatus, 186
   pilosus, 186
Asteraceae characteristics, 375

bamboo, Japanese, 290
Barbarea vulgaris, **200**, 368
barberry, Japanese, 11, **42**, 369
barnyardgrass, **338**

Bartram, John, 114, 148, 164, 176, 250, 274, 320, 352
bedstraw
    catchweed, 306
    smooth, **306**
beggarticks, devil's, **156**
Berberis thunbergii, **42**, 369
Bermudagrass, 346
Berteroa incana, **202**
berberine, 42, 104
Betula
    nigra, **46**, 368, 373
    nigra 'Heritage,' 46
    pendula, 23, 48, 373
    populifolia, **48**, 373
Bidens frondosa, **156**
bindweed
    black, 282
    field, 224
    hedge, **224**, 368, 370
biodiversity, 15, 18, 19
birch
    European silver, 23, 48, 373
    gray, **48**, 373
    red, 46
    river, **46**, 368, 373
bittercress, hairy, **206**, 368
bittersweet
    American, 58
    Oriental, **58**, 368
blackberry
    common, **98**
    cutleaf or evergreen, 98
blackcap, 102
bluegrass
    annual, **356**
    Canada, 356
    Kentucky, **358**
    wood, 358
Boston ivy, 130
bouncing bet, **214**, 370, 372
box elder, **112**, 367, 373
box elder beetle, 112
Brassica kaber, 200
Brassicaceae characteristics, 375
brome
    downy, **330**
    Hungarian or smooth, 330
Bromus
    inermis, 330
    tectorum, **330**

Bryum
    argenteum, **30**
    caespiticium, 30
bryum, silky, 30
buckthorn
    common, **88**, 369, 370, 374
    glossy, **86**, 369
buckwheat, climbing, **282**
burcucumber, 228
burdock, common, **152**, 370, 371
burning bush, **60**
butter and eggs, 274
buttercup
    bulbous, **298**
    creeping, **300**
    tall, 298

Calystegia sepium, **224**, 368, 370
campion
    bladder, 218
    white, **218**
canarygrass, reed, **350**
Capsella bursa-pastoris, **204**, 368, 370, 371
Cardamine hirsuta, **206**, 368
cardenolides, 142
carpetweed, 234, **264**, 288, 296, 367, 368
carrot, wild, 22, **138**
Carya cordiformis, 74, 373
Caryophyllaceae characteristics, 375
catalpa
    common, 50
    hardy or northern, **50**, 368, 373
Catalpa
    bignonioides, 50
    speciosa, **50**, 368, 373
catbrier, 362
cattail
    common, **364**
    narrowleaf, 364
celandine
    greater, **270**, 370, 371
    lesser, 300
Celastrus
    orbiculatus, **58**, 368
    scandens, 58
Celtis occidentalis, **52**, 373
Centaurea
    biebersteinii, 158
    maculosa, 158
    stoebe, 10, **158**, 367

Cerastium
    fontanum, **212**
    vulgatum, 212
Cerasus serotina, 92
Chamaesyce maculata, 234
charlock, 200
cheatgrass, 330
cheeses, 262
Chelidonium majus, **270**, 370, 371
Chenopodium album, **136**, 367, 371
Chernobyl (Ukraine), 154
cherry
    black, **92**, 374
    choke, 92, 374
chickweed, 371
    common, 82, **222**
    mouseear, **212**
chicory, 10, 12, 22, **160**, 367, 372
Chrysanthemum leucanthemum,
        176
chufa, 324
Cichorium
    endivia, 160
    intybus, 10, 12, 22, **160**, 367, 372
cinquefoil
    oldfield, 302
    rough, 304
    silvery, **302**
    sulfur, **304**
Cirsium
    arvense, 162
    vulgare, **162**
cleavers, 306
Clematis
    terniflora, **84**
    virginiana, 84
clematis, sweet autumn, **84**
climate change, 3, 16–17, 18, 20–21,
        368–369
clotbur, 152
clover
    alsike, 244, 372
    rabbitfoot, 246
    red, **244**, 372
    white, **246**
    white sweet, **240**
    yellow sweet, 10, 240
cocklebur, **194**
coltsfoot, 190, 370
Commelina communis, **322**
compass plant, 174

Convolvulus
    arvensis, 224
    sepium, 224
Conyza canadensis, 168
copperleaf
    rhombic, **230**
    slender, 230
    Virginia, 230
corktree, Amur, **104**, 374
compass plant, 174
Coronilla varia, 242
cosmopolitan urban meadow, 22, 372
cottonwood, eastern, **106**, 373
couch grass, 340, 346
coumarin, 240
crabapple
    Siberian, 90
    Toringo, 90
crabgrass, 367
    hairy or large, 336
    smooth, **336**
crownvetch, **242**
cucumber, wild, **228**
Cynanchum
    louisae, 144
    nigrum, 144
Cynodon dactylon, 346, 368
Cyperus
    esculentus, **324**
    strigosus, 324

Dactylis glomerata, **332**
daisy, oxeye, **176**, 372
dandelion, **190**, 367, 371
Darlington and Thurber, 38, 76, 80, 100,
        108, 120, 122, 138, 188, 208, 268, 270,
        272, 288, 294, 308, 332, 338
Darwin, Charles, 228, 260
Datura
    innoxia, 310
    stramonium, **310**
Daucus carota, 22, **138**, 370
dayflower, Asiatic, **322**
deadnettle, purple, 256
deer-tongue grass, **334**, 372
Dennstaedtia punctilobula, 32
dewberry, northern, **100**
Dichanthelium clandestine, **334**, 372
Digitaria, 368
    ischaemum, **336**, 367
    sanguinalis, 336

Dioscorides, 9, 370
disturbance, 3–6, 14–16, 368
dock
    broadleaf, 294
    curly, **294**, 371
dog-strangling vine, 144
dogbane
    hemp, **140**
    spreading, 140
Dutch elm disease, 112, 124, 126
Draba verna, 206
Dysphania ambrosioides, 136

Echinochloa crus-galli, **338**
Echinocystis lobata, **228**
ecosystem services, 18
Edison, Thomas, 180
Egler, Frank, 23–24
Elaeagnus
    angustifolia, 62, 373
    umbellata, **62**
Eleusine indica, 336
elm, 368
    American, 20, **124**, 368, 374
    Siberian, **126**, 374
Elymus repens, **340**
Elytrigia repens, 340
endive
    Belgium, 160
    curly, 160
epazote, 136
epigenetics, 6–7
Equisetum
    arvense, **34**
    hyemale, 34
Eragrostis, 368
    pectinacea, **342**, 367
    spectabilis, 342, 372
Erechtites hieracifolius, **164**
Erigeron
    annuus, **166**
    canadensis, **168**
Erophila verna, 206
escarole, 160
euonymus, winged, 60
Euonymus
    alatus, **60**
    europaeus, 60
    fortunei, 60
Eupatorium
    altissimum, 148

rugosum, 148
serotinum, 148
Euphorbia
    cyparissias, **232**
    maculata, **234**, 367, 368
Euphorbiaceae characteristics,
    376
Eurybia divaricata, 148, 186
evolution, 6–7

Fabaceae characteristics, 376
Fallopia
    convolvulus, 282
    japonica, 290
    scandens, **282**
fern
    hayscented, 32
    sensitive, **32**
fescue
    red, 344, 372
    tall, **344**
Festuca
    arundinacea, **344**
    elatior, 344
    rubra, 344, 372
Ficaria verna, 300
fireweed, **164**
fleabane, annual, **166**
fleur-de-lis, 326
foxtail
    giant or Chinese, 360
    green, **360**
    yellow, 360
Frangula alnus, **86**, 369
Frankia (bacteria), 44, 62
Fraxinus
    americana, 78, 373
    pensylvanica, **78**, 368, 373

Galeopsis tetrahit, **252**
galinsoga
    hairy, **170**
    smooth, 170
Galinsoga
    parviflora, 170
    quadriradiata, **170**
Galium
    aparine, 306
    mollugo, **306**
gallant soldier, 170
Garden of Eden, 162

garlic, wild, **320**
gill-over-the-ground, 254
glacier, urban, 5
Glechoma hederacea, **254**, 370, 371
Gleditsia triacanthos, **66**, 368, 373
globalization, 12–13
gobo, 152
goldenrod
    Canada, **180**
    early, 182, 372
    late, 180
    seaside, **182**, 372
    tall, 180
goosegrass, 336
grape
    'Concord,' 132
    fox, 132
    riverbank, **132**
Gray's Manual of Botany, 11
greenbrier, roundleaf, **362**
ground ivy, **254**, 370, 371
groundsel, common, **178**, 371

hackberry, **52**, 373
Hall, George Rogers, 36, 54
hawkweed
    field, 172
    mouse-eared, 172
    New England, 12, **172**
hay fever, 20, 150, 154
healall, **258**
heart's ease, 286, 318
heat-island effect, 16–17, 368, 369
hemp nettle, **252**
henbit, **256**
herbicide resistance, 24, 34, 134, 190, 246, 322
Hesperus matronalis, 202
hickory, bitternut, 74, 373
Hieracium
    caespitosum, 172
    canadense, 172
    lachenalii, 172
    pratense, 172
    subaudum, 12, **172**
Hogg, Thomas, 56, 58, 80, 84
honeysuckle, 11, 369
    Amur, 56
    Japanese, **54**, 368
    Morrow's, **56**
    Tatarian, 56

hops
    European, 210
    Japanese, **210**
horsetail, field, **34**
horseweed, **168**
Humulus
    japonicus, **210**
    lupulus, 210
    scandens, 210
Hypericum perforatum, **250**, 370, 371

Impatiens capensis, **196**
indigo bush, false, **64**
intaglio, 24
inulin, 160
invasive species, 3–4, 10, 12, 18, 21, 29
Ipomoea purpurea, 224
iris, yellow flag, 326, 370
Iris pseudacorus, **326**, 370
ivy
    Boston, 130
    ground, **254**, 370, 371
    poison, 20, **40**

jewelweed, **196**
jimson weed, **310**
Jonny jump-up, 318
Josselyn, John, 9, 371
Juglans nigra, **74**, 373
Juncus tenuis, **328**, 367
Junegrass, 358

Kenfield, Warren G., 24
kiwicha, 134
Klamath weed, 250
knapweed, spotted, 10, **158**, 367
knawel, **216**
knot-grass, 288
knotweed
    Bohemian, 290
    giant, 290
    Japanese, **290**
    prostrate, 234, 264, **288**, 296, 367, 371

Lactuca serriola, **174**
ladysthumb, **286**, 370, 371
    creeping, 286
    Oriental, 286
lambsquarters, common, **136**, 367, 371
Lamiaceae characteristics, 376

Lamium
  amplexicaule, **256**
  purpureum, 256
lawn, freedom vs. industrial, 24
leadwort, 64
Leonurus cardiaca, 252, 370
Lepidium
  campestre, 208
  virginicum, **208**
lettuce, prickly, **174**
Leucanthemum vulgare, **176**, 372
Ligustrum
  obtusifolium, **80**
  vulgare, 80
Linaria
  canadensis, 274
  vulgaris, **274**, 371
Linneaus, 27
Liquidambar styraciflua, 368
locust
  black, 11, **68**, 367, 368, 373
  honey, **66**, 368, 373
Lolium arundinacea, 344
Lonicera
  japonica, **54**, 368
  maackii, 56
  morrowii, **56**
  tatarica, 56
  species, 369
loosestrife, purple, **260**
Lotus corniculatus, **236**, 372
lovegrass
  purple, 342, 372
  tufted, **342**, 367
lucerne, 238
Lychnis alba, 218
Lythrum salicaria, **260**

maintenance, horticultural, 14–15, 18, 22
mallow, common, **262**, 367, 371
Malus
  baccata, 90
  pumila, 90, 374
  sieboldii, 90
Malus species, **90**, 369
Malva
  neglecta, **262**, 367, 371
  rotundifolia, 262
maple
  Norway, **114**, 374
  red, 116, 373

silver, **116**, 367–368, 373
sugar, 114
sycamore, **116**, 373
Matricaria discoidea, 178
meadow
  cosmopolitan urban, 372
  European, 22
medic, black, **238**
Medicago
  lupulina, **238**
  sativa, 238, 370
Melilotus
  albus, **240**
  officinalis, 240
mercury, three-seeded, 230
Mexican tea, 136
Meyer, Frank, 94
Microstegium vimineum, 334
milkweed, common, **142**, 372
milk sickness (trembles), 148
mimosa, 66
Mollugo verticillata, **264**, 367, 368
monarch butterfly, 144, 182
morning glory, tall, 224
Morus alba, **76**, 368, 373
  variety *multicaulis*, 76
moss
  purple or fire, 30
  silvertip, **30**
motherwort, 252, 370
mugwort, **154**, 367
Muhlenbergia schreberi, **346**
mulberry, white or common, **76**, 368,
   373
mullein
  common, 12, **308**, 370
  moth, 308
mustard
  garlic, **198**, 368
  wild, 200

Native Americans (plant usage), 34, 38, 50,
   72, 106, 108, 112, 130, 132, 134, 152, 156,
   164, 166, 168, 180, 194, 196, 208, 228,
   234, 266, 268, 272, 278, 284, 312, 316,
   318, 354, 364
neophytes, 8–9, 19
nettle
  American stinging, **314**, 371
  European stinging, 314, 370
  hemp, **252**

New Amsterdam, 10
nightshade
  bittersweet, **122**
  Eastern black, **312**, 371
  European black, 312, 370
nimblewill, **346**
nitrogen-fixation, 62, 66, 68, 238
nitrogen deposition, 17, 198, 367
nutgrass, yellow, 324
nutsedge
  false, 324
  yellow, **324**
Nuttall, Thomas, 88
Nuttallanthus canadensis, 274

oak
  pin, **70**, 368, 373
  red, **72**, 374
Oenothera biennis, **266**
olive
  autumn, **62**
  Russian, 62, 373
Onoclea sensibilis, **32**
orchardgrass, **332**
Oxalis
  corniculata, 268
  stricta, **268**
oxalic acid, 268, 292

Panicum, 368
  dichotomiflorum, **348**
  virgatum, 348
panicum, fall, **348**
Parthenocissus
  quinquefolia, **130**
  tricuspidata, 130
Paulownia tomentosa, **82**, 373
pear
  'Bradford,' 94
  callery, 94, 373
pearlwort
  birdseye, 216
  trailing, 216
Penn, William, 358
pennycress, field, 208
pepperweed
  field, 208
  Virginia, **208**
Persicaria
  caespitosa, 286
  lapathifolia, **284**

  longiseta, 286
  maculosa, **286**
  pensylvanica, 284
Phalaris arundinacea, **350**
  'Picta,' 350
Phellodendron amurense, **104**, 374
Phleum pratense, **352**
Phragmites australis, 3, **354**
  subsp. *americanus*, 354
  subsp. *australis*, 354
  variety *gigantissima*, 355
Phytolacca americana, **272**
phytoremediation, 18, 34, 48, 110,
    136, 148, 154, 160, 174, 322, 326,
    350, 354
pigweed, 368
  prostrate, 134
  redroot, **134**
pilewort, 164, 300
Pilosella offinarium, 172
pineapple weed, 172
Plantago
  lanceolata, **276**
  major, **278**, 371
  rugelii, 278
plantain
  blackseed, 278
  broadleaf, **278**, 371
  buckhorn, **276**
Plymouth colony, 10, 188, 296
Poa
  annua, **356**
  compressa, 356
  nemoralis, 358
  pratensis, **358**
Poaceae characteristics, 376
poison ivy, 20, **40**, 196
pokeweed, **272**
Polygonaceae characteristics,
    376
Polygonum
  aviculare, **288**, 367, 371
  cuspidatum, 290
  lapathifolium, 284
  persicaria, 286
  scandens, 282
poplar
  carolina, 105
  eastern or necklace, 23, 106
  hybrid, 242
  white or silver, 108, 370, 373

Populus
  alba, 108, 370, 373
  deltoides, 23, **106**, 373
  nigra, 106
  tremuloides, 4, **108**, 373
  x *canadensis*, 106
porcelain berry, 368
Portulaca oleracea, **296**, 367, 368, 370,
    371
Potentilla
  argentea, **302**
  norvegica, 304
  recta, **304**
  simplex, 302
preadaptation, 6–7, 367–369
primrose, evening, **266**
princess tree, **82**, 373
privet
  border, **80**
  European, 80
Prunella vulgaris, **258**
Prunus
  serotina, **92**, 374
  virginiana, 92, 374
purslane, common, 234, 264, 288, **296**, 367,
    368, 370, 371
Pyrus calleryana, **94**, 373

quackgrass, **340**
Queen Anne's lace, 22, 138, 370
Quercus
  palustris, **70**, 368, 373
  rubra, **72**, 374
quinoa, 136

radicchio, 160
ragweed, common, 20, **150**
Ranunculaceae characteristics, 377
Ranunculus
  acris, 298
  bulbosus, **298**
  ficaria, 300
  repens, **300**
raspberry, black, **102**
reed, common, 3, **354**
restoration, ecological, 21, 24
resveratrol, 290
Reynoutria
  japonica, **290**
  sacchalinense, 290
  x *bohemica*, 290

Rhamnus
  cathartica, **88**, 369, 370, 374
  frangula, 86
Rhizobium, 64, 68, 238, 242, 244, 246,
    248
Rhus
  glabra, 38, 373
  radicans, 40
  typhina, **38**, 373
ribbon grass, 350
Robinia pseudoacacia, 11, 66, **68**, 367, 368,
    373
rocket
  dame's, 202
  yellow, **200**, 368
Rosa multiflora, **96**, 369
Rosaceae characteristics, 377
rose, multiflora, 11, **96**, 369
Rubus
  allegheniensis, **98**
  flagellaris, **100**
  laciniatus, 98
  occidentalis, **102**
  phoenicolasius, 102
Rumex
  acetosa, 292
  acetosella, **292**
  crispus, **294**, 371
  obtusifolius, 294
rush
  path, **328**, 367
  scouring, 34

Sagina
  decumbens, 216
  procumbens, 216
salicin, 106, 108, 110
Salix
  bebbiana, 110
  discolor, 110
  humilis, 110
  purpurea, 110
  viminalis, 110
  x *sepulcralis*, 110, 373
salsify
  meadow, **192**
  western, 192
  yellow, 192
sandspurry, red, **220**
Saponaria officinalis, **214**, 370, 372
  'Flore Pleno,' 214

saponins, 214, 218
Scleranthus annuus, **216**
Securigera varia, **242**
Sedum
   acre, **226**
   sarmentosum, 226
self-heal, 258
Senecio vulgaris, **178**, 371
Setaria, 368
   farberi, 360
   pumila, 360
   viridis, **360**
shade-tolerance ratings (trees), 373–374
Shakers, 40, 156, 164, 168, 254, 258, 292
Shakespeare, William, 152, 278, 318
shepherd's purse, **204**, 368, 370, 371
Sicyos angulatus, 228
Silene
   latifolia, **218**
   vulgaris, 218
silktree, 66, 368, 373
Sinapis arvensis, 200
smartweed
   pale, **284**
   Pennsylvania, 284
Smilax rotundifolia, **362**
snakeroot, white, **148**
soapwort, 214
Solanaceae characteristics, 377
Solanum
   dulcamara, **122**
   nigrum, 312, 370
   ptycanthum, **312**, 371
Solidago
   altissima, 180
   canadensis, **180**
   gigantea, 180
   juncea, 182, 372
   sempervirens, **182**, 372
Sonchus
   arvensis, 184
   oleraceus, **184**, 370, 371
sorrel, red, **292**
sourgrass, 268, 292
sowthistle
   annual, **184**, 370, 371
   perennial, 184
spear grass, 356
speedwell
   common, 280
   corn, **280**, 371

purslane, 280
   thymeleaf, 280
Spergularia rubra, **220**
spiderwort
   Ohio, 322
   Virginia, 322, 372
spindletree, European, 60
Spirn, Anne Whiston, 1
spontaneous urban vegetation, 1–2, 4–5,
   8–11, 13–15, 18, 19–24
spurge
   cypress, **232**
   spotted, **234**, 264, 288, 296, 367, 368
St. Johnswort, common, **250**, 370, 371
Stellaria media, **222**, 371
sticktights, 156
stiltgrass, Japanese, 334
stonecrop
   goldmoss, **226**
   stringy or trailing, 226
succession, 5, 15–16, 24
sumac
   smooth, 38, 373
   staghorn, **38**, 373
sustainability, 18, 22, 24
swallowwort
   black, **144**, 368
   pale, 144, 370
sweetgum, 368
switchgrass, 348
Symphyotrichum
   cordifolium, 186
   pilosum, 12, **186**, 372

Tanacetum vulgare, **188**, 371, 372
tansy, common, **188**, 371, 372
Taraxacum officinale, **190**, 367, 371
Taxus
   baccata, 36
   brevifolia, 36
   cuspidata, **36**, 374
   cuspidata 'Nana,' 36
thistle
   bull, **162**
   Canada, 162
   star, 158
Thlaspi arvense, 208
thorn apple, downy, 310
thoroughwort
   late, 148
   tall, 148

timothy, **352**
toadflax
    oldfield, 274
    yellow, **274**, 371
touch-me-not, 196
Toxicodendron radicans, 20, **40**
Tradescantia
    ohiensis, 322
    virginiana, 322, 372
Tragopogon
    dubius, 192
    porrifolius, 192
    pratensis, **192**
tree-of-heaven, 2, **120**, 367, 368, 373
trefoil, birdsfoot, **236**, 372
Trifolium
    arvense, 246
    hybridum, 244, 372
    pratense, **244**, 372
    repens, **246**
    repens 'Ladino,' 246
Tussilago farfara, 190, 370
Typha
    angustifolia, 364
    latifolia, **364**

Ulmus
    americana, 20, **124**, 368, 374
    pumila, **126**, 374
urban landscapes
    managed, 13–15
    remnant, 13–14
    ruderal, 13–14
urbanization, 8, 11–13, 16–17, 19–21
Urtica dioica, 314, 370, 371
    subsp. *gracilis*, **314**
    subsp. *procera*, 314

Velcro, 150
Verbascum
    blattaria, 308
    thapsus, **308**, 370
Verbena
    hastata, 316, 372
    urticifolia, **316**
Veronica
    arvensis, **280**, 371
    officinalis, 280
    peregrina, 280
    serpyllifolia, 280

vervain
    blue or swamp, 316, 372
    white, **316**
vetch, bird, **248**
Vicia cracca, **248**
virgin's bower, 84
Vintoxicum
    nigrum, **144**, 368
    rossicum, 144, 370
Viola
    sororia, **318**
    tricolor, 318
violet, common blue, **318**
Virginia creeper, **130**
Vitaceae characteristics, 377
Vitis labrusca, 132
    'Concord,' 132
Vitis riparia, **132**

wallpepper, 226
walnut, black, **74**, 373
weeds, 3–5
whitlow grass, 206
willow
    basket, 110
    purple osier, 110
    pussy, **110**
    weeping, 110, 373
wineberry, 102
winter cress, 200
wintercreeper, 60
woodbine, 130
Woodson, Robert, 1
woodsorrel
    creeping, 268
    yellow, **268**
wormseed, 136

Xanthium strumarium, **194**

yarrow, **146**, 370, 372
yew
    English, 36
    Japanese, **36**, 374
    Pacific, 36

zelkova, Japanese, 126
Zelkova serrata, 126, 374